Student Lab Manual

Symmetry, Shape, and Space
with The Geometer's Sketchpad®

L. Christine Kinsey
Canisius College

Teresa E. Moore
Ithaca College

Key College Publishing
Innovators in Higher Education

www.keycollege.com

L. Christine Kinsey
Department of Mathematics
Canisius College
Buffalo, NY 14208, USA
kinsey@canisius.edu

Teresa E. Moore
Department of Mathematics and
 Computer Science
Ithaca College
Ithaca, NY 14850, USA
moore@ithaca.edu

Key College Publishing was founded in 1999 as a division of Key Curriculum Press in cooperation with Springer-Verlag New York, Inc. It publishes innovative curriculum materials for undergraduate courses in mathematics, statistics, and mathematics and statistics education. For more information, visit us at www.keycollege.com.

Key College Publishing
1150 65th Street
Emeryville, CA 94608
www.keycollege.com
(510) 595-7000

Development Editor: Susan Minarcin
College Production Project Manager: Michele Julien
Copyeditor: Cathy Baehler
Production Director: Diana Jean Parks
Production Service: scrathgravel publishing services
Art and Design Coordinator: Kavitha Becker
Cover Designer: Paul Reed
Prepress: DeHart's Printing Services Corporation
Printer: DeHart's Printing Services Corporation

Executive Editor: Richard Bonacci
General Manager: Mike Simpson
Publisher: Steven Rasmussen

Printed in the United States of America
10 9 8 7 6 5 4 3 2 1   07 06 05 04 03

ISBN: 1-931914-13-3

To my mother, Marjorie W. Kinsey

To my parents, Lorraine W. and J. Richard Engel

Introduction

This lab manual is meant to accompany the text <u>Symmetry, Shape, and Space</u>. It is intended to supplement the primary text, rather than serve as a self-contained manual. The sections of the manual roughly parallel the chapters of <u>Symmetry, Shape, and Space</u>. The section topics from the first half of the text, dealing with two-dimensional phenomena, are essentially duplicated and so complement the text nicely. However, you will find that some sections in <u>Symmetry, Shape, and Space</u> correspond to two sections in the lab manual because we felt it would take more than one class period in a lab to cover the material. The later chapters of <u>Symmetry, Shape, and Space</u> on three- and four-dimensional geometry do not as easily lend themselves to the use of computers.

We are excited about this pairing of our text with the capabilities of *The Geometer's Sketchpad®*. The formal instruction on geometric constructions and operations from <u>Symmetry, Shape, and Space</u> can immediately be applied to computer constructions, extending and consolidating the student's understanding of the material. The dynamic nature of *The Geometer's Sketchpad* encourages the student to experiment and create many variations on each theme. The ease of construction and the interactive properties of the sketches allow the student to freely investigate geometric properties and to clarify points of interest.

<u>Symmetry, Shape, and Space</u> was consciously written to contain approximately three times as much material as can be covered in a single semester. Hence, there is a great deal of flexibility in designing a course. We have tried to retain this flexibility in the lab manual. Later sections generally do not refer explicitly to earlier constructions, though they do assume an increasing familiarity with the features and menu options available in *The Geometer's Sketchpad*. When a previous construction method is needed, we refer the user to the appropriate section.

Designing a Course Using <u>Symmetry, Shape, and Space</u> and the Lab Manual

A course using this manual to accompany <u>Symmetry, Shape, and Space</u> could be taught in several ways. A class might meet one day a week in a computer lab to do a section of this lab manual corresponding to the chapter from the text that they have been studying in class. Alternatively, a class might be held in a computer lab and alternate between explorations with paper and pencil and turning to *The Geometer's Sketchpad* when applicable or convenient. If a computer lab is not available to the students, the instructor might use this lab manual as a source of computer demonstrations to be projected for viewing by the class.

The demonstrations of this text are purposefully detailed. This allows the instructor or student to work through the examples with little room for error. The demonstrations also provide instruction on using the various tools and menus of *The Geometer's Sketchpad*. After using the demonstrations as an introduction to a topic, the reader can then proceed to the exercises and to experiment with variations on each theme.

The sections of this manual vary in length. While many are appropriate for a class period in a computer lab, others consist of short demonstrations suitable for in-class presentation. See the table on the following page to coordinate the chapters of the primary text with the sections of the lab manual.

Some of the exercises in this lab manual duplicate those of the primary text. Others are extensions of the material. Having students do both the paper and pencil constructions and the software constructions is valuable in many circumstances. Choosing between the two is reasonable in others. We leave it to the instructor to make these judgments. The notation [SSS x.y.z] in an exercise indicates that the exercise parallels the given exercise from <u>Symmetry, Shape, and Space</u>. Some of these exercises have been reworded or extended for use with *The Geometer's Sketchpad*.

Using *The Geometer's Sketchpad* with the Lab Manual

The instructions in this manual are for Version 4 of *The Geometer's Sketchpad*. Many of the activities can be done with earlier versions, but some specific statements such as "the line will change color" will not be true. We assume only very basic knowledge of *The Geometer's Sketchpad* to begin with. We have tried to make the instructions quite complete, especially in the earliest sections. It is suggested that all students begin with **Section 0: Warm-up Activity** and then do Section 1 and 3 before proceeding through the rest of the text. We have maintained a fairly explicit level of instruction throughout the manual as we do not expect people to proceed through the sections in a linear manner. We have found that the interface and menus of *The Geometer's Sketchpad* are so intuitive that most students pick up what they need to know with little formal instruction. We spend between five and ten minutes demonstrating how to use the **Toolbox** tools to construct points, lines, and circles, how to use the selection arrow to click on and select one or more objects, and how to pull down and choose menu options. Then we turn the students loose to play with the software. For a more formal introduction to the software, we refer the user to the Quick Reference Guide reprinted in the Appendix or to Section 6 of this lab manual. Section 6 explores the basic ruler and compass constructions on which the program is based. The instructions for the exercises in this section are very explicit and are generally matched to the construction capabilities of the software. To work directly from *The Geometer's Sketchpad*, try looking under **Help: Contents: Toolbox Reference** at the **Overview of the Toolbox** and under **Help: Contents: Menu Reference** for how to use a particular menu.

Finally, there are sketches that accompany many of the exercises and demonstrations in this manual. These are indicated with a ★ symbol. They can be used by instructors to present the activity to a class without having to recreate the sketch. They can also be used by students not comfortable with computers to explore an idea and/or to see how changes in some values change the result. These sketches are identified by title in a note following the demonstration or exercise to which they relate. All of these sketches are downloadable, and are available at http://www.keycollege.com/SSS. Many of the sketches use JavaSketchpad technology, which means they are dynamic and useful for demonstration purposes right on the Web.

Coordination of Symmetry, Shape, and Space, the Lab Manual, and the Sketches

The table on the following page gives correspondences between sections of this laboratory manual, chapters of Symmetry, Shape, and Space, and the accompanying sketches.

Acknowledgments

We would like to thank Susan Minarcin, Richard Bonacci, Mike Simpson, and Michele Julien at Key College Publishing for suggesting that we write this manual and for their support during the process. We are grateful for many helpful suggestions from the reviewers: Helen Gerretson, Ramin Naimi, and Joel Zeitlin. Steven Chanan at Key was very helpful when it came to the inner workings of *The Geometer's Sketchpad*. Thanks to Anne Draus and Scratchgravel Publishing Services for many of the commas and all of the semicolons. And of course we would like to thank Canisius College and Ithaca College and our colleagues for their support while we completed this project.

Contents

0. Warm-up Activity

It is very easy to construct something that looks like a square using only the **Segment** tool of *The Geometer's Sketchpad*®. However, this kind of drawing will not pass what the software creators call "the drag test." If you click and hold on one of the points or segments of your sketch and then move it, your figure will no longer be a square. Using more involved techniques and the menu options, you can create a square that will stay a square no matter how you move the points or change the segments. We describe one method of constructing a "draggable" square in great detail below with the goal of getting you used to working with *The Geometer's Sketchpad*. A more detailed introduction is available in <u>The Geometer's Sketchpad: Learning Guide</u> and through the **Help** menu. Also see the Quick Reference Guide in the Appendix.

Demonstration: Construct a Square

1. Construct a line segment using the **Segment** tool. First, click on the line segment in the tool bar. Then, in the drawing window, press and hold while you drag the mouse. Release the mouse button and you have a line segment. The segment is highlighted, meaning it is currently "selected."

2. Click on the **Selection Arrow** tool. Click on empty space in the sketch to deselect the line segment (and anything else that may be selected). Then click on each endpoint of the segment. A colored ring will cover an object when it is selected. Choose **Show Label** from the **Display** menu. This will put the letters A and B next to the points you selected, with A next to the point you selected first and B next to the second.

3. Click on the **Circle** tool, and put the cursor on endpoint B of your segment. (The point will get a colored circle around it when the cursor is over it.) Hold down the mouse button and drag over to endpoint A. You will see a circle with increasing radius as you move the mouse. When the second endpoint is highlighted, release the mouse button. You now have a circle of radius AB centered at B. Note that the last object constructed is automatically selected. You can deselect it by using the **Selection Arrow** tool and clicking either on the object or on empty space.

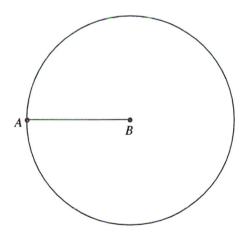

4. Click on the **Selection Arrow** tool. Select the line segment AB and the point B by clicking on them. Then choose **Perpendicular Line** from the **Construct** menu. This constructs the line through point B that is perpendicular to line segment AB. If **Perpendicular Line** does not appear in black, you cannot choose it because you do not have the correct combination of objects selected. (Quite likely, you still have the circle from the last step selected.) Click on empty space, and try again.

1

5. Click on the **Point** tool. Put the cursor at one of the points where the circle intersects the perpendicular line. When you are at the right point, both the line and the circle will change color. (Any object is highlighted when the cursor is over it.) Clicking when both are highlighted will create the point of intersection.

6. Choose **Show Label** from the **Display** menu. This should label the point C. Note: The point should be selected as a result of the construction. If it is not, click on the **Selection Arrow** tool. Select the intersection point and then choose **Display: Show Label**.

7. Click on the **Segment** tool. Put the cursor on point B (which should change color), and press and drag to point C. When point C changes color, you are in the right place. Release your mouse button. This constructs line segment BC, which is the second side of your square. Notice that line BC will change color as you construct segment BC whenever the cursor is on the line. This means that if you release the mouse button when the whole line is highlighted, you are constructing a segment from B to another point on line BC and not to C itself.

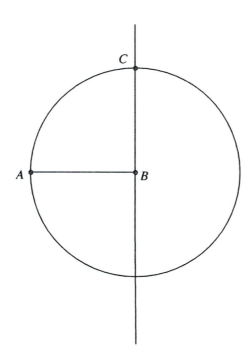

8. Following the procedure of Step 4, construct the line that goes through point C and is perpendicular to segment BC and the line that goes through point A and is perpendicular to segment AB. (Alternately, you can select point C and segment AB, or point A and segment BC, and then choose **Parallel Line** from the **Construct** menu.)

9. Click on the **Point** tool. Put the cursor at the point where the two lines constructed in Step 8 intersect. When you are at the right point, both lines will change colors. Click here to create the point of intersection.

10. Choose **Show Label** from the **Display** menu. This should label the point D. Note: The point should be selected as a result of the construction. If it is not, click on the **Selection Arrow** tool. Select the intersection point and then choose **Display: Show Label**.

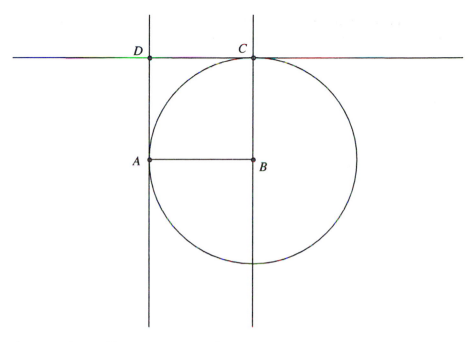

11. Following the procedure of Step 7, construct line segments *AD* and *CD*.

12. Using the **Selection Arrow** tool, select the circle and the lines. Select a line rather than a segment by clicking on the part of the line that is not part of the square. (In general, if there are several objects on top of one another, you can select different ones by clicking on the region until the object you want is highlighted. In this case, clicking on segment *AD* will alternate between selecting the segment and the line.)

13. Choose **Hide Path Objects** from the **Display** menu. The wording of this option will change depending on what type or types of objects you have selected.

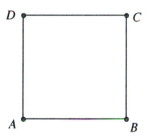

What you should see now is a square labeled *ABCD*. If you are missing sides of the square, you either forgot to construct them or you hid them by mistake. You can choose **Display: Show All Hidden** and try again, or you can reconstruct the segments using the **Segment** tool. Once you have a square, press on the mouse button and hold on any of the line segments or on point *C* or *D* and move your cursor. The square should move but not change shape. If you move *A* or *B*, the size of the figure should change but it should remain a square. The original segment can be changed, but the rest of the construction is tied to this segment. Your square passes "the drag test."

1. Area

Companion to Chapter 1.1 of <u>Symmetry, Shape, and Space</u>

Demonstration 1: Construct a Parallelogram [★ See the sketch Area.gsp: Parallelogram.]

We will use *The Geometer's Sketchpad* first to construct a parallelogram and then to verify that the area of a parallelogram does not depend on the length of the slanted "diagonal" side. We will construct a parallelogram and show that sliding one side along the parallel line on which it rests changes the shape but not the area of the figure. That is, as long as the base of the parallelogram stays the same, the top segment can be anywhere on the opposite parallel line without changing the area of the figure.

1. Construct a line segment AB, either by using the **Segment** tool from the tool bar or by placing two points and then using the **Segment** option in the **Construct** menu.

2. Select the endpoints of the segment, and choose **Show Label** from the **Display** menu. This should label the points A and B, depending on the order in which you selected the points.

3. Construct (using the **Point** tool from the tool bar) a point not on the line segment. We will call this point O here and on the following sketch, but we suggest that you do not label it. Note: If you use **Display: Show Label**, this point will automatically be labelled C on your sketch. You can change that by double-clicking on the label. A pop-up box, **Properties of Point C**, will appear with two tabs, **Object** and **Label**. If necessary, click on the **Label** tab, fill in the **Label** blank with an O, and check the **Show Label** box. Then click the **OK** button. *The Geometer's Sketchpad* labels things sequentially, and asking it to **Show Label** will automatically label the next point with the next letter (in this case, D); even if you change the label of this point, the following point will be labelled with the following letter (E). You may then need to adjust the notation in the following instructions to match your sketch or rename each point as you label it. Very few sketches are done without erasing at least one point so most students get used to adjusting or changing the notation quickly.

4. Select your line segment and point. From the **Construct** menu, choose **Parallel Line**. This should construct the line that runs through the point and parallel to the line segment.

5. Construct a line segment beginning at endpoint A of your original segment and ending at a point on the parallel line you constructed in Step 4. Release the mouse button when the line changes color to ensure that the end of the segment is on the line. Do not connect A to O, the third point you constructed; choose a new point on the parallel line. Moving O will allow you to move the parallel line. We want this line to be a fixed distance from AB so we can see that the figure's area does not change as the location of the top moves along the parallel line.

6. Select the new endpoint and choose **Display: Show Label**, which should label the point C.

7. Select the new line segment AC and the endpoint B of the original segment. Choose **Parallel Line** from the **Construct** menu.

8. Select the two lines and choose **Intersection** from the **Construct** menu. (If **Intersection** is not an option, you may have too many items selected. Click the selection arrow on empty space to clear it and reselect only the two lines.)

9. Select the intersection point (it is automatically selected in the construction process) and use **Display: Show Label** to name it D. By now, you should have a parallelogram and some extra lines, as shown in the following figure on the left.

10. Construct the line segments BD and CD.

11. Select the line through BD (not the segment) by clicking on the line at a point not on the segment, and select the point O you constructed to place the original parallel line. Then choose **Hide Objects** from the **Display** menu. You should have the figure shown below on the right.

Now, play with your figure to see what moves and to what effect, by dragging it with your mouse. If you select A or B, you should be able to change the whole drawing, but it will still remain a parallelogram. If you select the segment AB, you can change the height of the figure, but the parallel line CD will not move. If you choose line segments AC, BD, CD, or point D, the figure will move without changing shape.

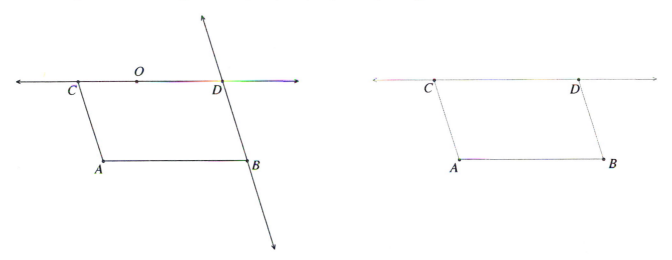

The point that demonstrates the independence of the area of a parallelogram and the length of the diagonal side is point C. If you move point C, the base and the height of the parallelogram will not change. To see that the area does not change:

12. Select—in order going around the figure either clockwise or counterclockwise—$ABDC$ (as labeled in the illustration) and choose **Quadrilateral Interior** from the **Construct** menu. (If the interior does not look like the parallelogram, check the order in which you chose your points and try again.)

13. With the interior highlighted, choose **Area** from the **Measure** menu.

14. Move A, B, or AB to see that the area changes as you change the width or height of the parallelogram.

15. Move point C, which is restricted to the parallel line, to see that area does not change if you change the shape of the figure without changing the width or height. The length of AC may be increased indefinitely but the area cannot change.

Demonstration 2: Verify That the Area of a Parallelogram Is Base Times Height

Construct a parallelogram as in Demonstration 1. The height of a parallelogram is the perpendicular distance between the parallel sides.

1. Select point A and line CD. Do not select only the line segment. Choose **Perpendicular Line** from the **Construct** menu.

2. Select the new line and the line CD. Choose **Intersection** from the **Construct** menu.

3. Select the new point (this is done automatically), and use **Show Label** from the **Display** menu to name it E.

4. Construct the line segment AE. This is the height of your parallelogram.

5. Hide the long line CD and the perpendicular long line AE, using **Display: Hide Lines**.

6. Select line segment AE and choose **Length** from the **Measure** menu. Do the same for segments AB and AC.

7. Select **Calculate** from the **Measure** menu. First, click on $m(AB)$, and you will see it appear in the calculator window. Next, hit the multiply button (marked with a $*$), and then click on $m(AC)$. Now, click **OK** on the calculator, and the result of the computation will appear on your sketch. This is the formula, $m(AB) * m(AC)$, that most students want to use for the area of a parallelogram.

8. Select **Calculate** from the **Measure** menu. Select $m(AB) * m(AE)$ and click OK. (If you do not have the area of the parallelogram calculated, follow Steps 12 and 13 in Demonstration 1 to construct the interior of the quadrilateral and measure its area.) As you move point C, you should notice that the value of $m(AB) * m(AC)$ changes. However, $m(AB) * m(AE)$ is constant and equal to the computed area.

▷ **Exercise 1.** Following the examples of Demonstrations 1 and 2, verify that the area of a triangle depends on the lengths of its base and height—not on the length of the other two sides—and that the area is equal to $\frac{1}{2}$ base times height. Your final sketch should have a triangle with a line parallel to the base and through the remaining vertex. Moving the vertex along the parallel should not change the triangle's area. Construct an altitude and demonstrate that Area $= \frac{1}{2}bh$. (If moving the vertex on the parallel line moves the line, review Steps 3–5 of Demonstration 1. Try constructing a line segment, a point, and a parallel line before you construct your triangle in order to maintain a constant height.)

▷ **Exercise 2.** [★ Area.gsp: Circle] Verify that the area of a circle is equal to πr^2 and that the circumference is equal to $2\pi r$, where r is the radius. There are several ways to see this. First, construct a circle. Then, measure the area, the circumference, and the radius. Using the **Calculate** feature, calculate the area divided by the radius and also the area divided by the square of the radius. The second value should look like π. Now, change the radius of the circle. The values for area, radius, and area divided by radius will all change. The value for area divided by the square of the radius will remain constant. Thus, area is proportional to the square of the radius. To verify the proportionality constant, calculate area divided by πr^2. The value should equal 1 for any circle. (You can find π under the **Values** button on the calculator. You can also change the number of digits displayed by using the **Edit: Preferences: Units** menu.)

▷ **Exercise 3.** Perform similar calculations to those in Exercise 2 to verify that the circumference of a circle is proportional to the radius but not to the square of the radius and that the constant of proportionality is 2π.

When asked about a trapezoid, many people will think of an isosceles trapezoid. Then, when they are asked to find the area, they begin by forming a rectangle inside with two equivalent triangles on the ends, such as Figure (a) below. Even if the trapezoid is not isosceles, most people will make the non-parallel sides slope away from each other, such as Figure (b). Few people consider the possibility illustrated in Figure (c).

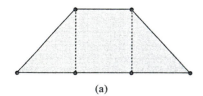
(a)

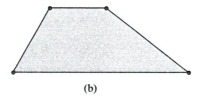
(b)

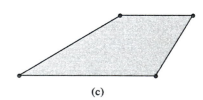
(c)

▷ **Exercise 4.** [★ Area.gsp: Trapezoid #1] Draw a trapezoid similar to Figure (b) on the previous page. Then, verify the area formula by breaking the figure into a rectangle and two triangles, and computing the sum of their areas. We include the numerical values for our drawing; yours will differ.

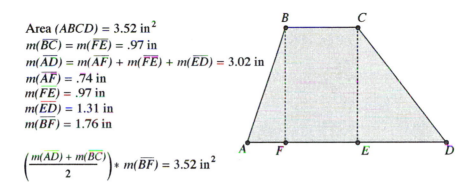

Area $(ABCD) = 3.52$ in^2
$m(\overline{BC}) = m(\overline{FE}) = .97$ in
$m(\overline{AD}) = m(\overline{AF}) + m(\overline{FE}) + m(\overline{ED}) = 3.02$ in
$m(\overline{AF}) = .74$ in
$m(\overline{FE}) = .97$ in
$m(\overline{ED}) = 1.31$ in
$m(\overline{BF}) = 1.76$ in

$\left(\dfrac{m(\overline{AD}) + m(\overline{BC})}{2}\right) * m(\overline{BF}) = 3.52$ in^2

▷ **Exercise 5.** Try the same approach that you used in Exercise 4 for figures similar to Figure (c) (repeated from the previous page) and Figure (d) below.

(c)   (d)

▷ **Exercise 6.** [★ Area.gsp: Trapezoid #2] Now, break a trapezoid into two triangles by connecting two non-adjacent vertices, and verify that the area formula is Area $= \frac{1}{2}b_1 h + \frac{1}{2}b_2 h = \frac{1}{2}(b_1 + b_2)h$.

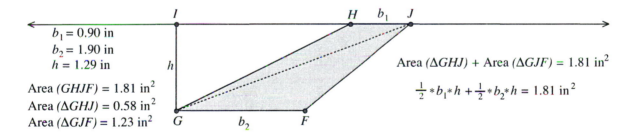

$b_1 = 0.90$ in
$b_2 = 1.90$ in
$h = 1.29$ in

Area $(GHJF) = 1.81$ in^2
Area $(\triangle GHJ) = 0.58$ in^2
Area $(\triangle GJF) = 1.23$ in^2

Area $(\triangle GHJ)$ + Area $(\triangle GJF) = 1.81$ in^2

$\frac{1}{2} * b_1 * h + \frac{1}{2} * b_2 * h = 1.81$ in^2

▷ **Exercise 7.** Give a different argument for the formula for the area of a trapezoid by showing that two copies of the trapezoid can be joined to form a parallelogram.

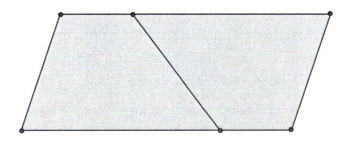

2. Tangrams

Companion to Chapter 1.2 of <u>Symmetry, Shape, and Space</u>

The tangram puzzle allows us an opportunity to introduce more of the construction tools, and to investigate different ways of getting the same result. You may want to work through Section 0: Warm-up Activity and some of the demonstrations and sketches that came with *The Geometer's Sketchpad* before doing this section.

Demonstration: Creating the Tangram Pieces [★ Tangrams.gsp: Tangram square]

We will construct the tangrams shown on page 15 of <u>Symmetry, Shape and Space</u> by constructing a large square and cutting it into the correct shapes. If you need directions on how to create a square, see Section 0: Warm-up Activity. Once you have a square labelled $ABCD$, the following directions will give the correct subdivision into the tangram pieces:

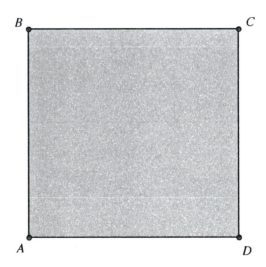

1. Construct the diagonal line segment AC from the lower left to the upper right corner.

2. Select this new line segment AC and choose **Midpoint** from the **Construct** menu to construct point E.

3. Construct the line segment BE from the upper left corner B to the midpoint E. You now have the two big triangles.

4. Construct the line segments AE and CE from the lower left A and upper right C corners to the midpoint E and construct the midpoints F and G for each of these segments.

5. Select the diagonal line segment AC (or AE) and the midpoint F. Choose **Perpendicular Line** from the **Construct** menu.

6. Select the perpendicular line you just constructed and the bottom of the square. Then choose **Intersection** from the **Construct** menu to create point H.

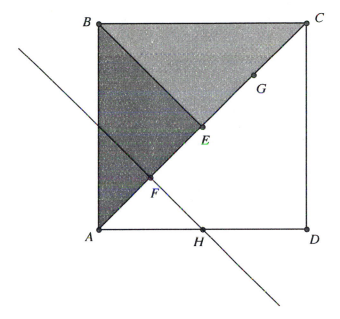

7. Construct the line segment between the midpoint F and the intersection point H.

8. Select the perpendicular line by clicking on it outside the square. Then choose **Hide Perpendicular Line** from the **Display** menu. Note that if the **Display** menu has **Hide Straight Objects** instead of **Hide Perpendicular Line**, you probably have both the line and line segment selected. Choosing this option hides the segment you just created. Go back and click on empty space to deselect everything. Then click on just the line and try again.

9. Select H and line segment FH. Then choose **Construct: Perpendicular**.

10. Select your newest line and the right side of the square, line segment CD. Choose **Construct: Intersection** to create point I.

11. Construct line segment HI. Then select the perpendicular line and choose **Display: Hide Perpendicular Line**.

12. Select line segment HI and choose **Construct: Midpoint** to create point J.

13. Construct line segments EJ and GJ.

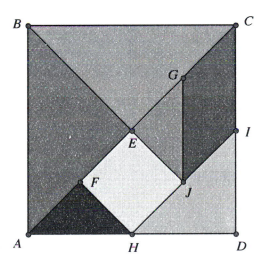

This finishes the construction of the tangram picture. In later exercises, you will need to form interiors of the shapes. This is done by selecting the corners around a shape in order and choosing **Interior** _____

from the **Construct** menu. The blank will read "triangle" or "quadrilateral" depending on the particular shape you have selected. You can make the shapes different colors by selecting an interior, choosing **Color** from the **Display** menu, and clicking on the color you want.

▷ **Exercise 1.** The instructions given above represent only one of many ways to construct the tangram puzzle. Find a different sequence of steps, and write careful instructions for your construction method.

▷ **Exercise 2.** Now construct a small square and rotate it 45° to make a diamond shape. Construct the rest of the tangram shapes by adding to this square instead of by cutting up a large square.

From either of your tangram sketches, create the **Interior** of each piece from the **Construct** menu, and make the pieces different colors. If you copy and paste the various interiors, then they can be moved around without reference to the other pieces. Use the **Rotate Arrow** tool (hold down the **Selection Arrow** tool and slide the mouse to the right to find this) to turn the pieces in place. When you copy and paste the interiors, the endpoints of the line segments used to construct the square will get copied, too. When you are done copying pieces, simply select all the extraneous points and choose **Hide points** from the **Display** menu. Now you are ready to start fitting pieces together. [★ Tangrams.gsp: Tangram pieces]

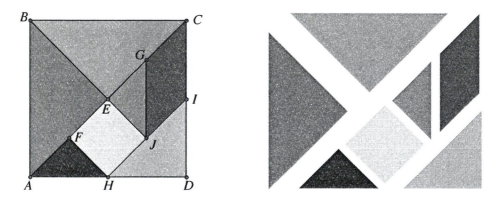

▷ **Exercise 3.** [SSS 1.2.3] Assemble the tangram pieces to form each of the 12 convex polygons shown below using *The Geometer's Sketchpad*.

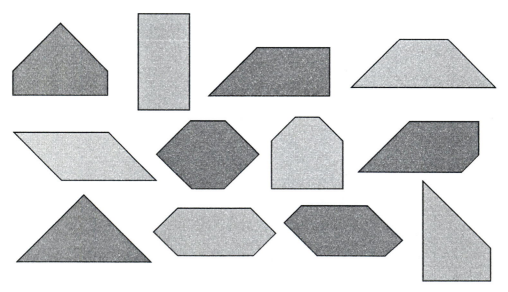

For the exercise above and any other time you try to solve a tangram puzzle with *The Geometer's Sketchpad*, you will need to switch back and forth frequently between the **Rotate Arrow** to turn the pieces and the regular **Selection Arrow** to move them. This can be tedious, so you may want to consider printing your sketch from the demonstration, cutting it apart, and working the old-fashioned way. When you first rotate a piece, *Sketchpad*^{TM} will choose a rotation point (the point on the drawing closest to the center of the screen) if you do not. This point will be the rotation point for the whole diagram. If a piece rotates off your viewing window, either enlarge the window or scroll up or down to find it, switch arrows to move the piece closer to the rotation point, and then rotate it further if necessary. Remember that the pieces must fit edge to edge with lengths matching exactly. Below is one possible solution to the first shape in Exercise 3. [★ Tangrams.gsp: Solution]

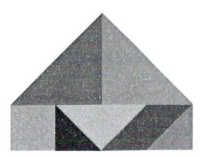

3. Polygons

Companion to Chapter 1.2 of <u>Symmetry, Shape, and Space</u>

▷ **Exercise 1.** Sketch a pentagon, a hexagon, a heptagon, and an octagon. You do not need to try to make them regular polygons. Divide each figure into the fewest possible number of triangles by connecting only existing vertices. At each vertex angle, choose the three points that describe the angle in order, and use the **Measure: Angle** menu to find the measure of the angle. Click on **Measure: Calculate** and a pop-up screen will appear. Compute the sum of the vertex angles for each of your polygons. Clicking on the angle measure will make it appear in the **Calculate** screen. The figure and the computations should lead you to the correct formula for Exercises 1.2.5–7 of <u>Symmetry, Shape and Space</u>.

▷ **Exercise 2.** The number of triangles you found in Exercise 1 above should be $(n - 2)$ for an n-sided polygon. If you let the lines dividing your polygons cross over each other, you may get a division into shapes other than triangles. If you introduce new vertices in the interior of your polygon, you can get many more triangles. If you only use existing vertices but allow the lines to cross, what is the maximum number of pieces you can get when dividing the polygons? For example, consider a pentagon. You can get as few as three pieces. However, if you connect each vertex to all the other vertices, you can get the picture below with 11 pieces—not all of them triangles.

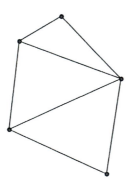 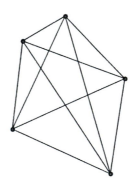

We will next give three methods for constructing regular polygons, which will be common figures in many of the chapters that follow. When asked to construct a regular polygon, many students will try to space points equally around a circle and then connect the dots. The first method given is based on this idea. However, this method gives only an approximation of a regular polygon. Moreover, the result is not stable: moving one of the points will destroy the regularity of the figure. The next two methods use the power of *The Geometer's Sketchpad* and the **Transform** menu to construct precise figures, which are stable. We will illustrate the three methods by constructing pentagons.

Demonstration: Constructing Regular Polygons: Method 1 [★ Polygons.gsp]

From Exercise 1.2.11 of <u>Symmetry, Shape, and Space</u>, you know that the interior angle of a regular pentagon is 72°. We will use this information to construct a regular pentagon.

1. Construct a circle using the **Circle** tool. Label the center B.
2. Construct a line segment from the center B of the circle to a point A on the edge of the circle.
3. Construct a second line segment from B to another point C on the edge.
4. Measure the angle $\angle ABC$ between these segments by selecting (in order) point A, point B, and point C and then choosing **Angle** from the **Measure** menu.
5. Move the second edge point C until the angle $\angle ABC$ is 72°. You are unlikely to get precisely 72°, so just get as close as you can.
6. Now construct a third line segment BD from the center to the rim of the circle, and make the angle $\angle ABD$ as close to 72° as you can.

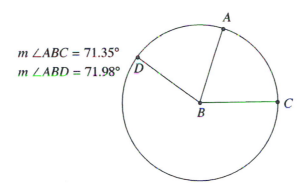

$$m \angle ABC = 71.35°$$
$$m \angle ABD = 71.98°$$

7. Do the same with a fourth and fifth segment. Be sure to measure all five angles: Small discrepancies in the first four can result in a large error in the fifth, in which case you must go back and adjust all the earlier angles.
8. When all the interior angles measure approximately 72°, construct line segments along the edge vertices to form a regular pentagon.
9. Now select the circle and interior line segments, and choose **Hide** from the **Display** menu.

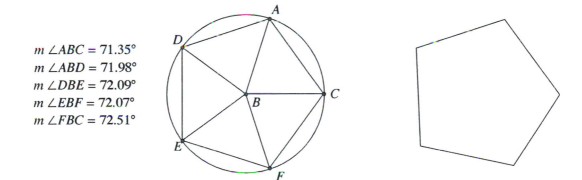

$$m \angle ABC = 71.35°$$
$$m \angle ABD = 71.98°$$
$$m \angle DBE = 72.09°$$
$$m \angle EBF = 72.07°$$
$$m \angle FBC = 72.51°$$

Note that this figure is an approximation to a regular pentagon. Using this method, it is almost impossible to get all of the angles to equal exactly 72°.

▷ **Exercise 3.** Following Method 1 outlined above and using your answers to Exercises 1.2.12 and 13 of <u>Symmetry, Shape, and Space</u>, construct a regular octagon and a regular dodecagon.

Demonstration: Constructing Regular Polygons: Method 2

1. First, construct two points A and B.

2. Select point A and choose **Transform: Mark Center** (or double-click on point A: It will flash, indicating that it has been chosen as a rotation center).

3. Now select point B and choose **Transform: Rotate**. A new screen labelled **Rotate** will appear. Make sure that the **Fixed Angle** option is checked and that it says **About Center A** at the bottom of the screen. Fill in the **Angle** blank with 72° and click on the **Rotate** button. A new point will appear at the correct angle.

B

A

4. Since the new point is already highlighted and we do not want to change the center of rotation, use the **Transform: Rotate** option to repeat the process until you have five points radiating from the center at A, all at 72° angles.

5. Construct line segments along the edge vertices to form a regular pentagon.

6. Now select the center point A, and choose **Hide Point** from the **Display** menu to get the picture below on the right. Click on the label at B and hide that also.

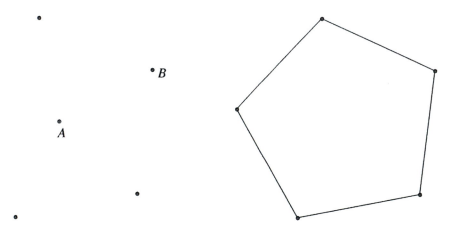

Dragging a side will move the figure. Dragging an exterior vertex will resize and rotate the figure without changing the shape. Note that one advantage to this method is that if one starts, for example, with two points at distance 1 apart, then the polygon formed will be inscribed in a circle of radius 1. One can thus create a family of polygons all inscribed in the same circle.

▷ **Exercise 4.** Following Method 2 outlined above and using your answers to Exercises 1.2.12 and 13 of Symmetry, Shape, and Space, construct a regular octagon and a regular dodecagon.

Demonstration: Constructing Regular Polygons: Method 3

Here is another way to construct a regular pentagon. We know that the vertex angle ought to be 108°.

1. First, draw a line segment AB.

14

2. Select point A and choose **Transform: Mark Center**.

3. Now select the line segment AB and the point B, and use the **Transform: Rotate** command. Again, make sure that the **Fixed Angle** option is checked and that **About Center A** appears at the bottom of the screen. Fill in the **Angle** blank with 108°, and click on the **Rotate** button. A new line segment AB' appears at a 108° angle from AB.

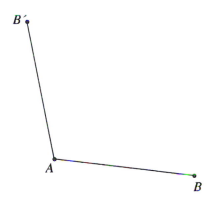

4. Now select point B', and use the **Transform: Mark Center** command.

5. Select the segment AB' and the point A, and choose **Transform: Rotate** to generate line segment $B'A'$.

6. Mark A' as the next rotation center, and rotate line segment $B'A'$ and point B' around it to make segment $A'B''$.

7. Finally, draw the line segment $B''B$ to close the figure.

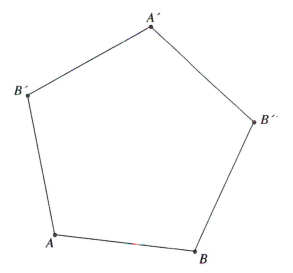

The advantage of this method is that if one starts, for example, with a line segment of length 1, then the polygon formed will have a uniform edge length of 1. One can thus create a family of polygons all with the same edge length.

▷ **Exercise 5.** Following Method 3 outlined above and using your answers to Exercises 1.2.12 and 13 of Symmetry, Shape, and Space, construct a regular octagon and a regular dodecagon.

▷ **Exercise 6.** Which procedure for the pentagon requires the fewest mouse clicks?

4. Billiards

Companion to Chapter 2.1 of <u>Symmetry, Shape, and Space</u>

In this section we give three ways of drawing a rectangular billiard table and three ways of plotting the trajectory of a ball on the table. Each has advantages and disadvantages. The first methods for each part of the process of modeling billiard problems are the simplest but also the most limited, and the last methods are more difficult but in most cases give more satisfactory results.

In exploring these problems we introduce the **Graph** menu and some of its options that pertain to grids. To lay out a grid using *The Geometer's Sketchpad*: Under the **Graph** menu, choose **Grid Form: Square Grid**. Two points are shown: one at the origin and one at position $(1,0)$, which determines the scale for the grid. If you would like a larger grid, drag the point at position $(1, 0)$ to the right, although if you drag this point too far, the software will instead introduce grid lines at half intervals.

Demonstration: Constructing a Billiard Table: Method 1 [★ Billiard tables.gsp: Method 1]

We wish to place the corner points of our billiard table precisely. One way to do this uses the command **Plot Points** under the **Graph** menu. A pop-up menu will appear. Make sure that the option **Coordinates: Rectangular** is checked, and then fill in the x- and y-coordinates of the point you wish to plot. This point is then fixed in place and cannot be moved. For example, we will draw a 5 by 3 table.

1. Plot points at the positions $(3,0)$, $(0,5)$, and $(3,5)$ using the **Graph: Plot Points** menu.

2. Draw the outer line segments connecting these corner points using the **Segment** tool or **Construct: Segment**.

3. Use the **Display: Line Width: Thick** and **Display: Color** commands to thicken and color these edges.

4. Choose the four corners of the rectangular table in order, and use the **Construct: Quadrilateral Interior** to color the table itself.

The advantage and disadvantage to this method is that it is rigid: Nothing you do (except deleting something) is going to change the shape of this table. You can change the scale shown along the axes by dragging the point $(1,0)$, but that is all you can do.

Demonstration: Constructing a Billiard Table: Method 2 [★ Billiard tables.gsp: Method 2]

As in Table Method 1, lay out a square grid. To lay out the corner points, click on the option **Graph: Snap Points**. This option will help ensure that any points you place will be located precisely at the integer grid points. You can, if you insist, place points anywhere you choose, but the points will act as if there is a magnetic attraction toward the intersections of the grid.

1. After choosing **Graph: Snap Points**, place points at $(3,0)$, $(0,5)$, and $(3,5)$ with the **Point** tool.

2. Draw the outer line segments connecting these corner points, and thicken and color these edges.

3. Construct and color the table itself.

The advantage of this method is that it is very easy. The disadvantage is that it is not rigid: Dragging a corner point will change the shape of the figure dramatically. This is fine if you want to play billiards on a trapezoidal table but rather a nuisance otherwise.

Demonstration: Constructing a Billiard Table: Method 3 [★ Billiard tables.gsp: Method 3]

As in Table Methods 1 and 2, lay out a square grid.

1. Using **Graph: Snap Points** and the **Point** tool, place a point at position $(3, 0)$.

2. Using the **Segment** tool, draw the horizontal line segment connecting this point to the origin.

3. Select the origin and this line segment, and use the **Construct: Perpendicular Line** command to draw a vertical line. This line will lie on top of the y-axis.

4. While it is still selected from the construction process, use **Construct: Point on Object** and then drag this new point to the position $(0, 5)$. It is important that this point be on the perpendicular line and not on the y-axis, since if it is at position $(0, 5)$ on the y-axis, dragging the point $(3, 0)$ on the x-axis will result in movement of this point to keep the table in the same proportions. We want to be able to rescale the two dimensions separately.

5. Select the point $(3, 0)$ and the horizontal line segment, and construct a second perpendicular line using **Construct: Perpendicular Line**.

6. Construct a line through the point $(0, 5)$ perpendicular to the vertical line through the origin.

7. Select the right-hand vertical line and the upper horizontal line, and use the **Construct: Intersection** command to create the upper right corner of the table.

8. Draw line segments from $(0, 0)$ to $(0, 5)$, from $(0, 5)$ to $(3, 5)$, and from $(3, 0)$ to $(3, 5)$.

9. Thicken and color these edges.

10. Construct and color the interior of the table.

11. Use **Display: Hide Lines** to hide the long horizontal and vertical lines.

The advantage to this method is that it is somewhat rigid but not completely so: You can drag the points at $(3, 0)$ and at $(0, 5)$ to new positions and the table will remain rectangular. (If dragging the point $(3, 0)$ makes the point at $(0, 5)$ also move, then select the $(0, 5)$ point and choose **Edit: Split Point from Axis**. Then move the point back into position.) Dragging $(0, 0)$ or $(3, 5)$ will just move the table and the axes around without changing size or shape. The disadvantage is that this method takes more steps than the other two.

All three grid methods will, in the end, give you a picture like the following:

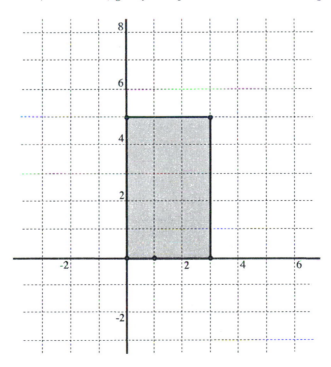

▷ **Exercise 1.** Build a 6 by 4 table by each of these three methods. Save your files, since you will be asked to use them in some of the following exercises.

Next, we will discuss three ways of drawing the trajectory for the ball. Our main emphasis will be on a ball that starts in the lower left corner, is shot at an angle of 45°, and bounces off the sides of the table until it reaches a corner where it stops. Start with a billiard table of your chosen dimensions. We will illustrate the processes with our 5 by 3 table. Place a guide point (using **Graph: Snap Points** and the **Point** tool) at position $(1, 1)$. The trajectory will always start at the origin and go through this point.

Demonstration: Drawing a Billiard Trajectory: Method 1

Again, this first method is the simplest and in some ways the most direct. Note that a 45° trajectory will cut each box of the grid that it goes through in half along the diagonal. Alternatively, think of the path of the ball as "up one–over one."

1. Draw a line segment (using the **Segment** tool) representing the first leg of the trajectory, from the lower left corner to the point where the ball impacts a side of the table.

2. Thicken your line segment and color it a different color from the colors used for the sides of the table and for the table itself.

3. Repeat the process following the bounce of the ball.

The advantage to this method is that it is very easy to do. The disadvantage is that any geometry involved (like figuring out the 45° angles) is done by you rather than by the computer. You are merely using *The Geometer's Sketchpad* as a drawing tool. This method is fine for 45° problems and Table Method 1, since you can't change the shape of the table, but inadequate for other angles and slopes or the other two methods of drawing the table. In Table Methods 2 and 3, dragging the corners of the table results in a trajectory with the same pattern of bounces but changes the angles.

▷ **Exercise 2.** Draw a 45° trajectory using Trajectory Method 1 on each of your 6 by 4 tables from Exercise 1, and try dragging the guide point at position $(1, 1)$ and then each of the corners of the table. Record your findings.

Demonstration: Drawing a Billiard Trajectory: Method 2

This method seems the most natural.

1. Draw a ray starting from the origin and going through the guide point at $(1, 1)$, using either the **Ray** tool or the **Construct: Ray** menu option.

2. Place a point (using either the **Point** tool or **Construct: Intersection**) where the ray intersects a side of the table.

3. Either double-click on that side (it should flash) or select it and use **Transform: Mark Mirror**.

4. Choose **Transform: Reflect**. A new ray will appear at the appropriate angle.

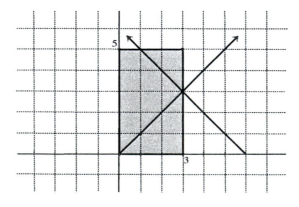

18

5. Repeat this process, marking the point of intersection with a side and then reflecting across that side, until the ball lands in a corner.

6. You will end up with a shaded rectangle representing the billiard table and a lot of rays crisscrossing (all at 45° angles to the sides of the table). The parts of these lines inside the rectangle form the path of the billiard ball. You should now go back and replace the rays by segments which remain inside the table, and then hide the rays. This creates a much less confusing picture.

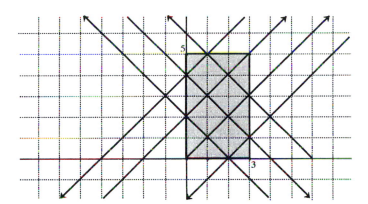 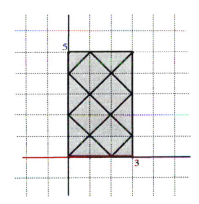

The advantage to this method is that it works quite naturally, as dragging the guide point will alter the trajectory correctly (as long as the ball bounces off the same sides as it did before). The disadvantage is that if you then alter the table dimensions or the slope of the trajectory, the trajectory may bounce off walls it doesn't even intersect. This method only works correctly if the new trajectory bounces off the same walls in the same order.

▷ **Exercise 3.** Draw a 45° trajectory using Trajectory Method 2 on each of your 6 by 4 tables from Exercise 1, and try dragging the guide point and then each of the corners of the table. Record your findings.

Demonstration: Drawing a Billiard Trajectory: Method 3 [★ Billiards. gsp]

1. Again, begin by drawing a line (use a line instead of a ray this time) from the origin through the guide point at (1, 1).

2. For a square table, this line will pass through the corner point opposite the origin. For any other table, the line will intersect one of the sides of the table. Create the point of intersection by choosing the trajectory line and the side of the table it intersects and using the **Construct: Intersection** command.

3. Select the intersection point and the table edge, and use **Construct: Perpendicular Line** to make a new line perpendicular to the side of the table.

4. Highlight this line and use the **Transform: Mark Mirror** command (or double-click on the line).

5. Select the diagonal line representing the trajectory of the ball, and use the **Transform: Reflect** command to generate the next section of the trajectory.

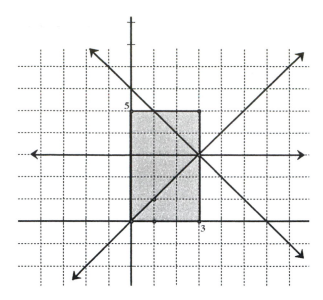

6. If this new line passes through any of the corner points, stop. Otherwise reflect the new line across the line perpendicular to the side of the table at the point of intersection with the trajectory line.

7. Repeat until the trajectory reaches a corner.

8. Finally, go back, replace all the long lines by line segments that stay within the rectangular table, and hide the lines.

You can drag the corners of the table to form new dimensions and to see the trajectory of the ball, but parts of the trajectory will disappear unless the trajectory bounces off the same edges. This is an advantage over Trajectory Method 2, which represented the ball bouncing off walls it didn't intersect. In fact, if you feel ambitious, you can drag the corner points and work out all possible reflections, so that all cases (up to however many bounces you feel like programming) are accurately displayed.

To investigate the trajectory of a ball shot at other angles, drag the guide point to a new position. (You may need to deselect **Graph: Snap Points**.) You can then observe the alterations in the trajectory that ensue. However, only the first reflection will be displayed accurately, unless the new trajectory bounces off the same walls in the same order as the original or you worked out alternate bouncing patterns as above.

▷ **Exercise 4.** Draw a 45° trajectory using Trajectory Method 3 on each of your 6 by 4 tables from Exercise 1, and try dragging the guide point and then each of the corners of the table. Record your findings.

▷ **Exercise 5.** Take your sketch from Exercise 4 and drag the point at $(4, 0)$ to position $(8, 0)$. The trajectory now hits the top wall first rather than the right-hand wall, so the reflection should disappear. Work out the new trajectory using Method 3. Now try dragging the guide point and each of the corners of the table. Record your findings.

The Geometer's Sketchpad illustrates the phenomenon of scaling very nicely. Drag the point at $(1, 0)$, which defines the grid, to the right. As you do this, the number of grid lines will first double, then quadruple, although the numbers along the axes will not change. In the following picture, while the numbers along the axes still indicate a 5 by 3 table, the number of grid lines gives 10 by 6. You can use **Display: Hide Axes** to make it appear to be a 10 by 6 table.

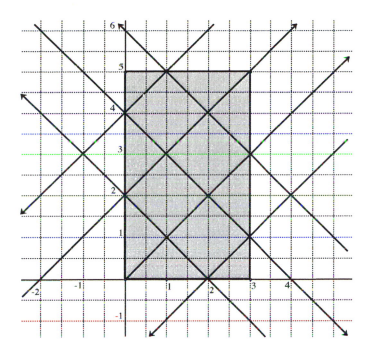

To investigate the unfolding of trajectories, start by drawing (and shading) the basic table and its corner points and edges as above. Use any of the table methods. Draw the ray that represents the first leg of the path of the ball. Highlight the table edge that this ray intersects, and choose **Transform: Mark Mirror** (or double-click on the edge). Then select the other table edges, the corners, and the interior of the rectangle, and use **Transform: Reflect**. Continue thus until the ray passes through a corner point.

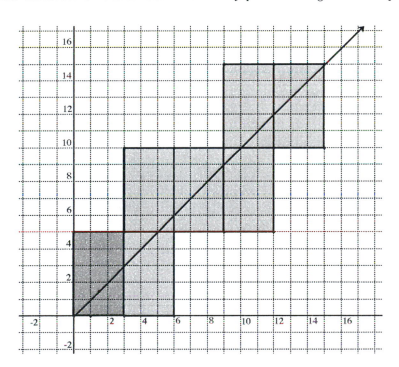

▷ **Exercise 6.**   Draw a billiard table using any of the table methods above but using a randomly placed guide point. Draw the trajectory using either Method 2 or 3. Use the **Measure: Angle** command to measure the angles the trajectory forms with the edges of the table. What do you notice?

For the problems below, use any method for the table and the trajectory, or better yet, do some of each, trying all combinations.

▷ **Exercise 7.** [SSS 2.1.2] Using *The Geometer's Sketchpad*, trace the path of a ball hit at a 45° angle on each of the tables below. Start in the lower left corner and follow each ball until it hits a corner.

(a) 6 by 3 table

(b) 9 by 3 table

(c) 12 by 3 table

(d) 6 by 6 table

(e) 9 by 6 table

(f) 10 by 8 table

(g) 8 by 6 table

(h) 7 by 3 table

(i) 15 by 9 table

▷ **Exercise 8.** [SSS 2.1.9] Trace the path of the ball on a table that is 5 by $2\frac{1}{2}$.

▷ **Exercise 9.** [SSS 2.1.10] Trace the path of the ball on a table that is $2\frac{1}{3}$ by 2.

▷ **Exercise 10.** [SSS 2.1.11] Trace the path of the ball on a table that is $1\frac{1}{3}$ by $1\frac{1}{2}$.

▷ **Exercise 11.** [SSS 2.1.14] (a) Using *The Geometer's Sketchpad*, trace the path of a ball hit at a slope of $\frac{1}{3}$ on a 6 by 3 table.

(b) Repeat for a 9 by 6 table.

▷ **Exercise 12.** [SSS 2.1.15] (a) Using *The Geometer's Sketchpad*, trace the path of a ball hit at a slope of 2 (think of 2 as $\frac{2}{1}$, so the ball goes 2 units up for every 1 unit to the right) on a 6 by 3 table.

(b) Repeat for a 10 by 6 table.

▷ **Exercise 13.** [SSS 2.1.17] Using *The Geometer's Sketchpad*, draw the unfolded trajectory for the following tables, assuming that the ball starts in the lower left-hand corner and is shot at an angle of 45°.

(a) 6 by 3

(b) 9 by 3

(c) 4 by 8

(d) 4 by 3

(e) 10 by 4

▷ **Exercise 14.** [SSS 2.1.20] Unfold the trajectory of a ball traveling along a path with

(a) slope $\frac{1}{3}$ on a 6 by 3 table

(b) slope 2 on a 6 by 4 table

▷ **Exercise 15.** **[SSS 2.1.21]** Plot the trajectory that the cue ball must travel to hit the black ball, if it bounces first off of the right-hand wall and then off of the left-hand wall. Use *The Geometer's Sketchpad* to draw the original table and situation of the two balls and to plan the trajectory.

▷ **Exercise 16.** **[SSS 2.1.22]** Figure out all possible two-cushion shots for the following configuration. Use *The Geometer's Sketchpad* to draw the original table and situation of the two balls and to plan each trajectory.

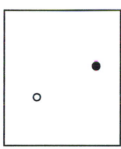

We'd like to thank Helen Gerretson for suggesting the following exercise.

▷ **Exercise 17.** You need to hike from point A to B pictured below, but you must stop at the river on the way. Figure out the most efficient path.

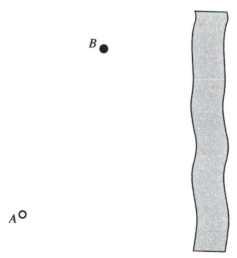

5. Celtic Knots

Companion to Chapter 2.2 of <u>Symmetry, Shape, and Space</u>

Demonstration: Drawing Celtic Knots [★ Celtic Knots.gsp: Grid of dots]

The Geometer's Sketchpad is not capable of doing the over-under weaving necessary in the formation of the celtic knot patterns. However, it is very useful for laying out the initial grid of dots and for drawing the guidelines. The final knot pattern can then be finished by hand.

1. To lay out the grid of dots, click on **Graph: Grid Form: Square Grid**.

2. Click on **Graph: Snap Points**.

3. Lay out a grid of dots of the required size for the knot you wish to draw by placing points at the grid intersections using the **Point** tool.

4. Go back and choose every alternate point and use the option **Display: Color** to make sure that the dots are laid out in alternating colors. For purposes of this discussion, we'll assume (since this will be printed in black and white) that you've chosen grey and black (though one might have preferred purple and pink). Be sure to begin your grid of dots with a grey dot in the upper left corner.

5. You can, at this point, use the commands **Graph: Hide Grid** to hide the grid itself, though the axes will still appear. Select the axes and choose **Display: Hide Axes** to make these disappear. You can at this point choose to print out your grid of dots and complete the celtic knot by hand.

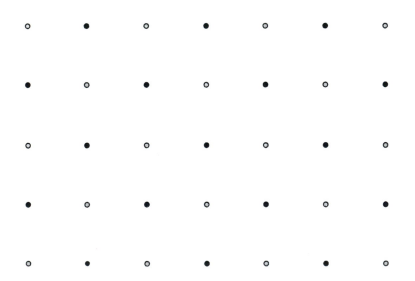

6. If you choose to draw the guidelines with the software, draw diagonal lines at an angle of 45° connecting the black dots only and omitting the corners where the guidelines would only be one diagonal long. [★ Celtic Knots.gsp: Guide lines]

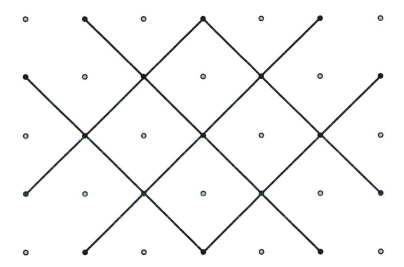

7. A further aid can be introduced by choosing to use guide circles at the corners and at the edges where the paths double back. First, deselect the option **Graph: Snap Points**.

8. Then draw a circle that fits nicely into the squares of the grid.

9. Copy and paste this circle to place copies as shown in the corners and along the edges. [★ Celtic Knots.gsp: Guide circles]

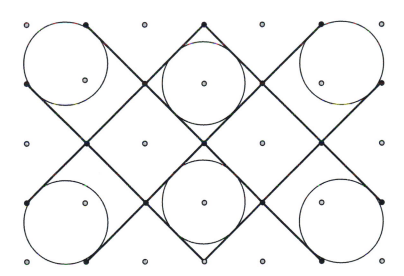

Use these guide circles to help round off the edges and corners in a uniform manner, stopping and starting your drawing lines just short of every other dot so that the over-under weaving pattern will appear in the final drawing.

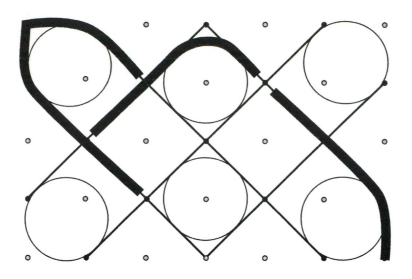

The drawing below was formed by using *Adobe® Illustrator®* over a grid of dots, lines, and circles generated by *The Geometer's Sketchpad*.

26

6. Ruler and Compass Constructions

Companion to Chapter 3.1 of <u>Symmetry, Shape, and Space</u>

The Geometer's Sketchpad was designed with traditional ruler and compass constructions in mind. Consequently, some of the exercises in this section will duplicate the exercises in <u>Symmetry, Shape, and Space</u>. You should develop a strong appreciation for how easily and precisely the software draws circles and matches lengths when compared to actually trying to draw figures with pencil and paper. However, the software has built into the **Construct** menu many of the constructions presented as exercises in this section. Therefore, we ask you to do most of the constructions twice: once using only the tool bar and again using the items of the **Construct** menu. The tool bar (located along the left side of the drawing window) contains tools for the basic ruler and compass objects: selecting objects; placing points; drawing circles; and drawing lines, segments, and rays. Once you learn to do the standard ruler and compass constructions using only the tool bar, you may use the options of the **Construct** menu which automates many of the constructions (such as locating the midpoint of a segment, constructing perpendiculars and parallels, constructing circles of a given radius, etc.). Be careful to follow the directions for each exercise describing which tools you are allowed to use.

    We give step-by-step instructions for performing the first ruler and compass construction using only the tool bar.

Demonstration: Ruler and Compass Construction 1: Perpendicular Bisector (*Construct the perpendicular bisector of a line segment.*) [★ Ruler and Compass.gsp: Construction 1]

1. Construct a line segment using the **Segment** tool. (That is, click on the line segment in the tool bar. In the drawing window, press and hold while you drag the mouse. Release the mouse button, and you will have a line segment. The segment is highlighted, meaning that it is currently selected.)

2. Click on the selection arrow tool. Click on open space to deselect the line segment (and anything else that may be selected). Then click on each endpoint of the segment. A colored ring will cover a point when it is selected. Choose **Show Label** from the **Display** menu. This will put the letters A and B next to the points you selected, with A next to the point you selected first and B next to the second.

3. Click on the **Circle** tool, and put the cursor on one endpoint of your segment. (The point will have a colored circle around it when the cursor is over it.) Hold down the mouse button and drag over to the other endpoint. You will see a circle with increasing radius as you move the mouse. When the second endpoint is highlighted, release the mouse button. You now have a circle of radius AB centered at A.

4. Following the procedure described in Step 3, construct a circle of radius AB centered at B.

5. Click on the **Point** tool, and move the cursor to a point where the circles intersect. When you are at the correct point, both circles will change color. Click here to insert the point of intersection. Then, repeat to form the other point of intersection of the two circles.

6. Click on the **Selection Arrow** tool. Select the intersection points, and choose **Show Label** from the **Display** menu. The points should be labeled C and D.

7. Select the **Segment** tool. Construct the line segment CD between the intersection points. Make sure the points are highlighted, not the circles, when you press and release the mouse button.

8. Select the **Point** tool and create the point where the line segments intersect. Label the point as in Step 6. (It should be point E.) Your picture should look like the following illustration.

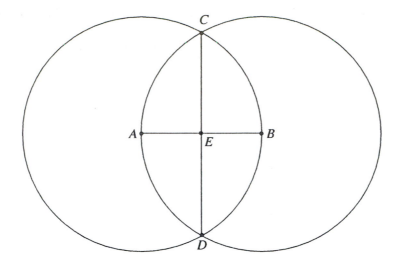

Point E is the midpoint of AB, and CD is the perpendicular bisector of AB. If you prefer to have a line as the perpendicular bisector instead of the line segment CD, in Step 7 click on the **Segment** tool and then slide to the right to choose the **Line** tool instead.

▷ **Exercise 1.** (a) Follow the directions above for constructing the perpendicular bisector of a line segment using only the tool bar.

(b) Next, use the menu options of *The Geometer's Sketchpad* to create the perpendicular bisector of a line segment. First, draw a line segment as above. Then select the segment and choose **Midpoint** from the **Construct** menu. Remember that the selection arrow tool must be used to select objects. Select the midpoint (it may still be selected from the construction) and the line segment. Then choose **Perpendicular Line** from the **Construct** menu.

By now, you should be pretty good at drawing points, circles, line segments, and lines and at selecting and labeling objects. We will not be as explicit in the following exercises regarding which tools you should be using.

Demonstration: Ruler and Compass Construction 2: Erect a Perpendicular to a Line through a Point (*Given a line and a point on that line, construct a line perpendicular to the given line through the given point.*) [★ Ruler and Compass.gsp: Construction 2]

1. Construct a line segment and a point on the line segment.

2. Select—in order—the left endpoint of the line segment, the right endpoint of the line segment, and the point you placed on the line segment. (Remember to click on empty space to deselect everything first.) Choose **Show Label** from the **Display** menu, so that the line segment is labelled AB and the point C.

3. Use the **Circle** tool to draw a circle centered at C and intersecting line segment AB at two points.

4. Use the **Point** tool to construct and label the points where the line segment and circle intersect as D and E.

5. Construct the perpendicular bisector of DE as above using only the tool bar. Note that this perpendicular bisector is a line through C and perpendicular to AB.

▷ **Exercise 2.** (a) Follow the directions above to erect a perpendicular to a line segment through a point on the segment using only the tool bar.

(b) Next, use the menu options of *The Geometer's Sketchpad* to erect a perpendicular. First, draw a line segment and a point on the segment as above. Then, select the segment and the point, and choose **Perpendicular Line** from the **Construct** menu.

Demonstration: Ruler and Compass Construction 3: Drop a Perpendicular to a Line from a Point (*Given a line and a point not on the line, construct a perpendicular line from the point to the line.*) [★ Ruler and Compass.gsp: Construction 3]

1. Construct a point C and a line AB. You will need to change the **Segment** tool to the **Line** tool to get the required setup. Remember that the point C should not be on the line AB.

2. Construct a circle centered at C and intersecting AB at two points, D and E.

3. Construct the perpendicular bisector of DE as in the first demonstration.

▷ **Exercise 3.** (a) Follow the directions above to drop a perpendicular to a line from a point off the line using only the tool bar.

(b) Next, use the menu options of *The Geometer's Sketchpad* to drop a perpendicular. First, draw a line and a point off the line as above. Then, select the segment and the point, and choose **Perpendicular Line** from the **Construct** menu.

Demonstration: Ruler and Compass Construction 4: Bisect an Angle (*Given an angle, construct a line which bisects that angle.*) [★ Ruler and Compass.gsp: Construction 4]

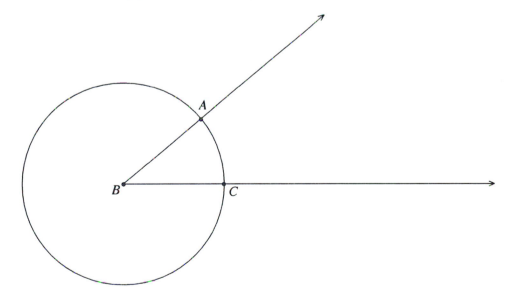

1. Construct an angle using either line segments or rays.

2. Construct a circle centered at the vertex B of your angle. Then construct the intersection points between the circle and each side of the angle. Label these points A and C.

3. Construct the circle centered at A with radius AC.

4. Construct the circle centered at C with radius CA.

5. Construct D, a point of intersection of the two circles of Steps 3 and 4.

6. Construct the line segment BD. This line bisects the angle $\angle ABC$.

▷ **Exercise 4.** (a) Follow the directions above to bisect an angle using only the tool bar.

(b) Next, use the menu options of *The Geometer's Sketchpad* to bisect an angle. First, draw the angle as above. Then, select the three angle points in order, and choose **Angle Bisector** from the **Construct** menu. Be sure to select the vertex of the angle as the second point. What happens if you select the points on the angle in a different order?

One of the nicer things about modern compasses is that you can use them to measure the length of a line segment or the radius of a circle and then transfer that length to another point. In particular, given a line segment and a point, it is easy to make a line segment starting at the point with the same length as the given segment. The compasses used in Euclid's time could not do this. They were made from two sticks of equal length with little holes drilled at one end and tied together with a bit of string. As soon as you picked them up, they collapsed so they could not be used to transfer lengths. The next construction will show you how to transfer a length using the collapsing compass of *The Geometer's Sketchpad*. We will then go over how to use the menu options to avoid repeating this complicated procedure.

▷ **Exercise 5.** First, construct a line segment and a point. The goal is to create a line segment starting at the new point with length equal to the length of the existing segment. Try to do this on your own using only the tools—not the menu options.

Demonstration: Floppy Compass [★ Floppy Compass.gsp]

If you have trouble with the preceding exercise, here is the procedure from Euclid's <u>Elements</u> (Proposition I.2) that describes how to transfer lengths using a floppy compass.

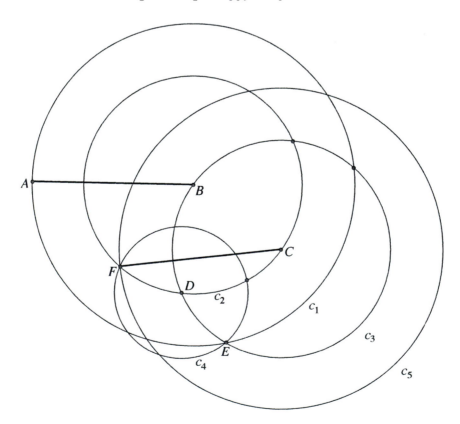

1. Construct a line segment and label the endpoints A and B. Construct a separate point and label it C. We will assume C is closer to B than to A for our picture, although it does not matter.

2. Construct the circle centered at B with radius BA. Use **Show Label** to name this circle c_1.

3. Construct the circles centered at B with radius BC and at C with radius CB. Label these circles c_2 and c_3 respectively.

4. Construct one of the points of intersection of circles c_2 and c_3. Label it D.

5. Construct the point where c_1 and c_3 intersect. Label it E.

6. Construct the circle centered at D with radius DE. Label it c_4.

7. Construct the point where c_2 and c_4 intersect. Label it F.

8. Construct the circle centered at C with radius CF. Label it c_5. The length of CF is equal to the length of AB, so any line segment beginning at C and ending on c_5 will have the desired length.

Demonstration: Fixable Compass

Now that you know how to do the construction by hand, we will use the **Construct** menu options to copy lengths easily. This models the modern fixable compass.

1. Construct a line segment and a point, and select both objects.

2. Choose **Circle by Center + Radius** from the **Construct** menu.

3. Create a line segment between the center of the circle and any point on the circle. The length of this new segment is equal to the length of the original.

▷ **Exercise 6.** Construct a line segment AB and a point C not on the line. Construct a new line segment with one endpoint at C and with length equal to AB using the **Construct** menu as above.

You can also copy and paste line segments, but there is no way to tell when the endpoint of a copied segment matches exactly with your point.

Demonstration: Ruler and Compass Construction 5: Copy an Angle (*Given an angle and a line segment, construct an angle on the given line segment congruent to the given angle.*) [★ Ruler and Compass.gsp: Construction 5]

First construct an angle $\angle ABC$ and a line segment DE. Then follow the steps below:

1. Use the **Construct: Circle by Center + Radius** command to draw a circle with center D and radius equal to BC. Label the point where the circle intersects DE as F.

2. Draw a circle with center D and radius AB.

3. Draw a circle with center F and radius AC. (You will need to construct line segment AC first.)

4. These latter two circles intersect at two points. Label one of them as G and construct the line segment DG.

5. The angle $\angle GDF$ is equal to $\angle ABC$.

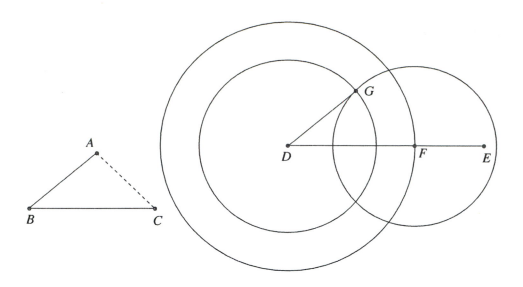

▷ **Exercise 7.** Draw an angle ∠*ABC* and place a point *D*. Construct an angle with vertex at *D* that is equal to ∠*ABC*.

Demonstration: Constructing Parallel Lines

The construction of copying an angle is used in the ruler and compass method for constructing parallel lines. Again, the software has this construction built into the menu options. We construct parallel lines first using only the tool bar and then using the menu commands.

1. Construct a line segment *AB* and a point *C* not on *AB*. We wish to construct a line through *C* that is parallel to *AB*.

2. Construct the ray from *A* through *C*.

3. Construct a point *D* on the ray so that *C* is between *A* and *D*.

4. Use the steps given in the previous demonstration to copy angle ∠*CAB* onto segment *CD*. After hiding all the extraneous objects, you should get a picture like the following:

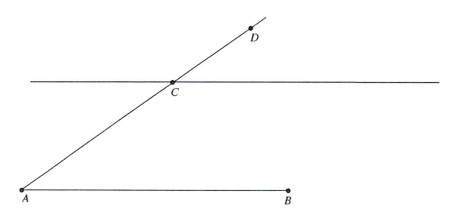

▷ **Exercise 8.** (a) Follow the directions above to construct a line through *C* parallel to *AB*.

(b) Construct a line through *C* parallel to *AB* using the **Construct** menu: Simply select the line *AB* and the point *C*, and use **Construct: Parallel Line**.

For the remainder of this lab manual, feel free to use any of the **Construct** menu commands, rather than restricting yourself to only the tool bar.

7. Constructibility

Companion to Chapter 3.1 of <u>Symmetry, Shape, and Space</u>

In this section, we will use *The Geometer's Sketchpad* to parallel the discussion of constructibility in <u>Symmetry, Shape, and Space</u>. We will study constructible lengths, angles, and polygons. Note that many lengths, angles, and polygons are constructible using the full power of *The Geometer's Sketchpad* that are not constructible by traditional ruler and compass. For example, using Constructing Regular Polygons Method 2 or 3 from Section 3 of this text (rotating the vertices by the appropriate angle) would allow one to construct a regular heptagon or nonagon, neither of which is constructible by euclidean tools. However, anything that is constructible by ruler and compass is also constructible by *The Geometer's Sketchpad* using only the tool bar and the **Construct** menu, and vice versa. Therefore, you may use any of those tools and commands in this section, but not those of the **Transform** menu.

We will first recall an additional ruler and compass construction from <u>Symmetry, Shape, and Space</u>. We will illustrate the procedure by dividing AB into six pieces.

Demonstration: Ruler and Compass Construction 6: Partition a Line Segment (*Given a line segment, construct points that divide the line segment into n equal pieces.*) [★ Ruler and Compass.gsp: Construction 6]

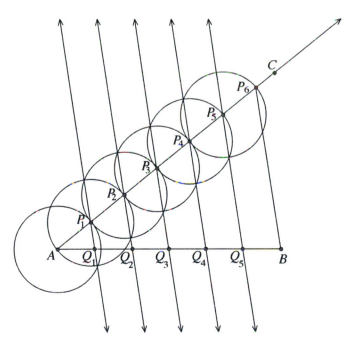

1. Let AB be the given line segment.
2. Draw a ray AC that does not coincide with line AB.
3. Set your compass on a arbitrary radius and draw a circle centered at A.
4. The intersection of this circle and the ray AC is called point P_1.

5. Using the radius AP_1, draw a circle with center P_1, and let P_2 be the intersection of this circle with AC as shown.

6. Again, using the same radius $AP_1 = P_1P_2$, draw a circle with center P_2, and let P_3 be the intersection of this circle with AC as shown.

7. Continue thus to find points $P_1, P_2, \ldots, P_n$ equally spaced along AC.

8. Draw the line segment BP_n.

9. Construct a line parallel to BP_n through each of the points $P_1, P_2, \ldots$. The line through P_i parallel to BP_n will intersect the line AB at a point we will label Q_i.

10. The points $Q_1, Q_2, \ldots, Q_{n-1}$ divide AB into n equal pieces.

▷ **Exercise 1. [SSS 3.1.7]** Construct a line segment, and use the steps from Ruler and Compass Construction 6 to divide it into three equal pieces.

While this is very time-consuming by hand, it is rather quick using the **Construct: Parallel Line** command of *The Geometer's Sketchpad*. Therefore, we suggest you try dividing a line segment into 10 pieces just to see how easy things can be.

▷ **Exercise 2. [SSS 3.1.8]** Construct a line segment. Construct a segment $\frac{4}{3}$ the length of the original, using only the tool bar and the items of the **Construct** menu. You should use the **Measure: Length** and **Measure: Calculate** options to confirm that you have the correct length.

The previous two exercises contain the basic techniques that will allow you to construct line segments of any rational length. That square roots of rational numbers are also constructible follows from those techniques and from the Pythagorean theorem. We begin by giving a technique to generate the square roots of integers.

▷ **Exercise 3.** To construct the Pythagorean spiral, begin by constructing a line segment AB. Find its length using the **Measure: Length** menu, and drag point B until it is 1 inch long. (You may have to open the **Edit: Preferences: Units** menu to change from centimeters to inches.) Construct the line through A perpendicular to AB, and draw a circle centered at A with radius $AB = 1$ inch. Let C be the point where the circle and the perpendicular line intersect. Draw the line segments AC and CB, and hide the long line and the circle. By the Pythagorean Theorem, $BC = \sqrt{2}$. Now construct another line through B and perpendicular to BC. Draw a circle centered at B with radius 1 inch. Label the intersection of the line and the circle as B' and note that $CB' = \sqrt{3}$. Continue constructing the spiral of right triangles, each with one side of length 1 inch (note $AB = BB' = B'B'' = 1$), until you have a hypotenuse of length $\sqrt{17}$. [★ Pythagorean spiral.gsp]

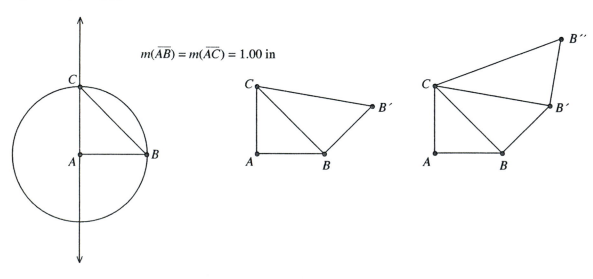

$m(\overline{AB}) = m(\overline{AC}) = 1.00$ in

▷ **Exercise 4.** [**SSS 3.1.10**] Draw a line segment. Construct, using only the tool bar and the **Construct** menu, another line segment that is $(1 + \sqrt{2})$ times the length of the original.

The next group of exercises explores the constructible angles. For these, you are only to use the tool bar and the **Construct** menu. For Exercises 7 and 8, you will need Ruler and Compass Construction 5.

▷ **Exercise 5.** Construct a right angle. Bisect this angle and measure one of the two new angles formed. Bisect this and measure one of the new angles. Repeat once more.

▷ **Exercise 6.** Construct an angle of 60°. Bisect this angle and measure one of the two new angles formed. Bisect this and measure one of the new angles. Repeat once more.

▷ **Exercise 7.** Construct an angle of 135°.

▷ **Exercise 8.** Construct an angle of 71.25°.

Finally, we address the regular polygons that are constructible by ruler and compass. Recall from Symmetry, Shape, and Space that these are precisely the regular n-sided polygons where all of the odd prime factors of n are distinct Fermat primes ($F_k = 2^{2^k} + 1$). Therefore, the list of constructible polygons begins: equilateral triangle, square, pentagon, hexagon, octagon, decagon, dodecagon, pentakaidecagon, etc.

▷ **Exercise 9.** [**SSS 3.1.11**] Using the tool bar and the **Construct** menu options, construct an equilateral triangle. Do not use any of the **Transform** options.

▷ **Exercise 10.** [**SSS 3.1.13**] Using the tool bar and the **Construct** menu options, construct a square. Do not use any of the **Transform** options.

▷ **Exercise 11.** [**SSS 3.1.14**] Using the tool bar and the **Construct** menu options, construct a regular hexagon. Do not use any of the **Transform** options.

▷ **Exercise 12.** [**SSS 3.1.15**] Using the tool bar and the **Construct** menu options, construct a regular octagon. Do not use any of the **Transform** options.

▷ **Exercise 13.** Using the tool bar and the **Construct** menu options, construct a regular dodecagon (12 sides). Do not use any of the **Transform** options.

▷ **Exercise 14.** Using the tool bar and the **Construct** menu options, construct a regular pentakaidecagon (15 sides). Do not use any of the **Transform** options.

8. The Pentagon and the Golden Ratio

Companion to Chapter 3.2 of <u>Symmetry, Shape, and Space</u>

Throughout western history, the golden ratio and rectangle have been considered the most visually pleasing proportion and shape. Often imbued with a mystical power, the ratio has been intentionally built into art and architecture. It also appears in many naturally occurring objects—or so many scholars will tell you. Whether beautiful, natural, or just a number, as mathematicians we are interested in methods of constructing the golden ratio.

The golden ratio is closely tied to the regular pentagon. As you know from Section 3.2 of <u>Symmetry, Shape, and Space</u>, if you can construct a regular pentagon, you can find a golden triangle. Conversely, if you can construct the golden ratio, you can construct a regular pentagon. Using the power of *The Geometer's Sketchpad*, constructing a regular pentagon is easy. Three methods are described in Section 3 of this manual. However, each of these methods used commands from the menu bar, either from **Transform** or **Measure**. If we allow only ruler and compass or only the tool bar and the **Construct** menu of *The Geometer's Sketchpad*, then constructing a regular pentagon requires first the construction of a golden ratio.

▷ **Exercise 1.** Use *The Geometer's Sketchpad* and the **Transform: Rotate** command to construct a regular pentagon. Construct the golden triangle inside the pentagon as shown below.

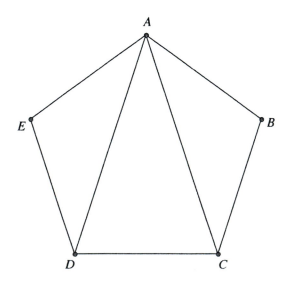

▷ **Exercise 2.** In your sketch from Exercise 1, hide the parts of the pentagon not related to the golden triangle. Construct a second golden triangle inside the first using the base of the original triangle as a side of the second, smaller one. Calculate the ratios $\frac{AC}{DC}$ and $\frac{DC}{FC}$. They should be equal to each other and to ϕ, the golden ratio.

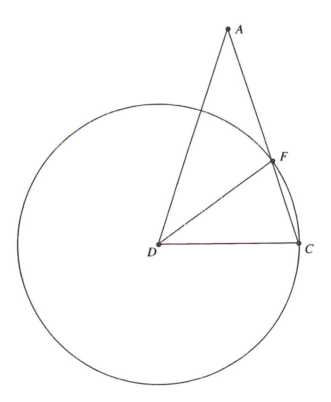

▷ **Exercise 3.** Following the procedure above, construct a third, fourth, and fifth triangle each nested in the previous one. Calculate the ratios of side to base to see that each consecutive triangle is indeed a golden triangle. [★ Golden ratio.gsp: Golden triangles]

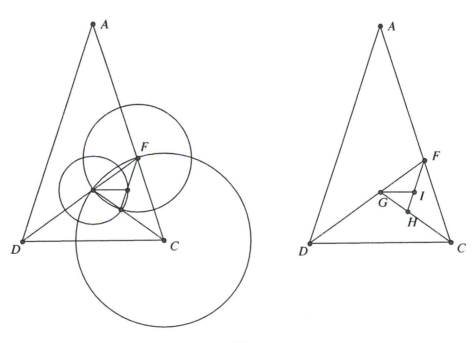

37

Constructing a perfectly regular pentagon without using the **Transform** menu of *The Geometer's Sketchpad* requires the construction of the golden ratio using the ruler and compass tools. We will also construct a golden rectangle along the way.

Demonstration: Ruler and Compass Construction 7: Construct a Line Segment to Represent the Golden Ratio (*Given a line segment, add a smaller segment so that the sum is to the original piece as the original is to the smaller.*) [★ Golden ratio.gsp: Golden ratio 1]

Follow the instructions below to construct line segments $AB = AD$, and then construct segment AF, so that $\frac{AF}{AD} = \frac{AD}{DF} = \phi$. Rectangle $ABGF$ is a golden rectangle. Note that we are following the steps for Ruler and Compass Construction 7 in <u>Symmetry, Shape, and Space</u>, but the labeling is slightly different to keep the construction rigid.

1. Construct a vertical line segment AB.
2. Construct a square with side AB. (See Section 0: Warm-up Activity.)

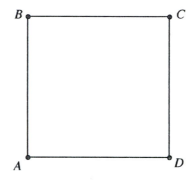

3. Construct the midpoint E of AD.
4. Construct the circle centered at E with radius EC.
5. Construct the ray AD, the intersection point F of the ray and the circle, and the line segment AF. The ratios $\frac{AF}{AD}$ and $\frac{AD}{DF}$ are equal to the golden ratio.

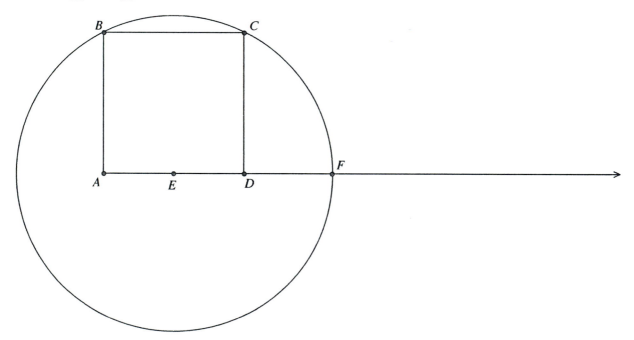

6. Construct the line through F perpendicular to AF, the ray BC, and the intersection point G of this line and this ray.

38

7. Construct line segments BG and FG.

8. Hide the circle and all rays and lines to leave golden rectangle $ABGF$.

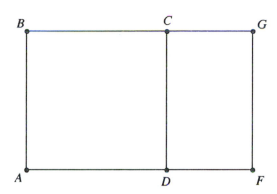

▷ **Exercise 4.** Use the procedure above to construct a line segment representing the golden ratio. When you have completed the drawing, moving B should automatically move D to keep the proportions as $\frac{AF}{AD} = \frac{AD}{DF} = \phi$.

Demonstration: Divide a Given Line Segment by the Golden Ratio [★ Golden ratio.gsp: Golden ratio 2]

Another simple way to divide a line segment by the golden ratio is illustrated by the following picture, which comes from 101 Project Ideas for The Geometer's Sketchpad.

$$\frac{AB}{AX} = \frac{AX}{XB} = \phi$$

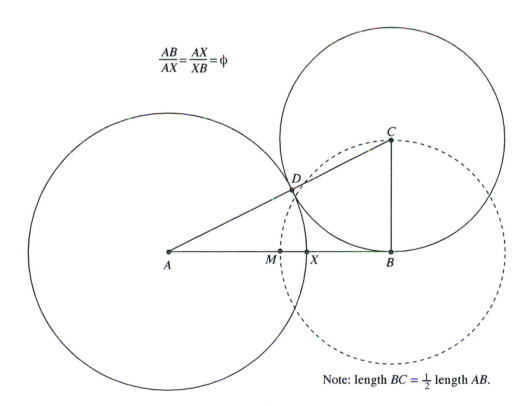

Note: length $BC = \frac{1}{2}$ length AB.

1. Construct a line segment AB.
2. Construct the midpoint M of AB.
3. Construct the circle centered at B with radius equal to MB.
4. Construct the line through B perpendicular to AB.
5. Construct the point of intersection C between the circle and the perpendicular line.
6. Construct line segment AC. Draw a circle centered at C that passes through B. Label the point D where this circle intersects the segment AC.
7. Construct a circle centered at A that passes through point D. Label the intersection X of this circle and the line segment AB.
8. Then hide the line BC, the segment AC, point D, the circles, and the midpoint of AB. The point X divides the segment AB by the golden ratio.

▷ **Exercise 5.** Use the procedure above to divide a line segment by the golden ratio. Note that the length of BC is half the length of AB. When you have completed the drawing, moving A or B should automatically move X to keep the proportions as $\frac{AB}{AX} = \frac{AX}{XB} = \phi$.

The process of constructing a golden ratio by either method can then be used to construct golden rectangles or golden spirals, often associated with the spiral form of the nautilus shell.

▷ **Exercise 6.** Copy A, B, X, and line segment AB of your sketch of Exercise 5 on a new page. (Moving A or B should still automatically move X to maintain the correct proportions.) Construct a golden rectangle on your segment.

More advanced users may be interested in constructing a **Custom Tool** that would subdivide a given line segment by the golden ratio. Consult <u>The Geometer's Sketchpad Reference Manual</u>. [★ Golden Ratio Tool. gsp]

▷ **Exercise 7.** Construct a golden rectangle $ABGF$ using either of the methods described above. Divide the shorter side of the rectangle $DCGF$ by the golden ratio and construct a smaller golden rectangle $CGHI$ inside it. [★ Golden ratio.gsp: Golden rectangles]

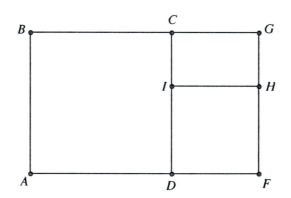

Note that in the drawing above $ABCD$ is a square, as is $DFHI$. Recall from <u>Symmetry, Shape, and Space</u> that the bigger part of the golden rectangle is a square and that removing this square leaves another golden rectangle. In other words, if you use Ruler and Compass Construction 7, you only need to create a square, instead of recreating the golden ratio to subdivide the smaller rectangle.

▷ **Exercise 8.** Building on the sketch you constructed for Exercise 7, continue to repeat the process of putting a golden rectangle (by forming a square) inside the smaller rectangle from the last round. Draw circles and then quarter circles inside the squares using the **Arc on Circle** command in the **Construct** menu to approximate the golden spiral. (Draw each circle. Select the circle and place the endpoints of the arc you want. Then go to the **Construct** menu. If the **Arc on Circle** option is not highlighted, try clearing your selections and starting again.) [★ Golden ratio.gsp: Golden spiral]

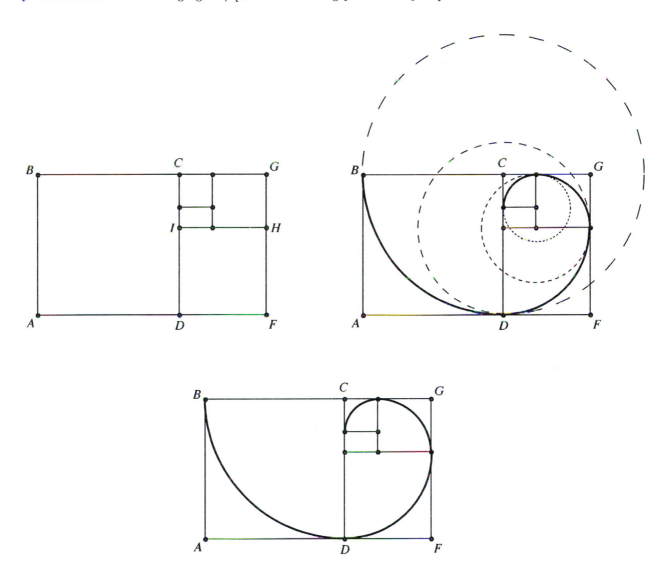

▷ **Exercise 9.** Construct a line segment AB. Construct a golden triangle with base AB and then a regular pentagon with AB as one side.

Please note that Exercise 9 is not as simple as it may seem at first glance. Until now, you have constructed golden ratios where the original line segment becomes the larger part of the ratio or the segment is broken into the ratio. Now you want the given piece to be the smaller part of the ratio. You should also note how much more efficient the pentagon constructions using *The Geometer's Sketchpad* **Transform** tools in Section 3 are when compared to constructions created with the traditional tools.

We end this section with a different construction of the regular pentagon inscribed in a circle by traditional ruler and compass methods. We need a circle with axes through the center to begin. Special thanks to Helen Gerretson for suggesting that we include this construction.

Demonstration: Inscribe a Regular Pentagon in a Circle

1. Construct a circle centered at A.

2. Construct a roughly horizontal line AB through the center of the circle using the **Line** tool (click and hold on the **Segment** tool and move the cursor over to the **Line** option before releasing).

3. Construct a line perpendicular to AB through A by selecting line AB and the point A and choosing **Construct: Perpendicular Line**.

4. Construct the points where these lines intersect the circle, either by selecting a line and the circle and choosing **Construct: Intersections**, or by using the **Point** tool to put a point where both the circle and the line change color. Label the intersection points N, S, E, and W (for north, south, east, and west).

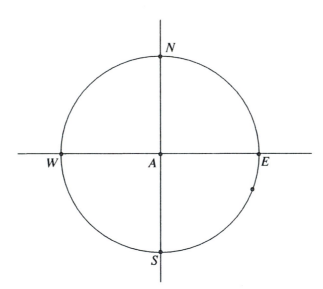

5. Construct line segment AW using the **Segment** tool. Select this segment and choose **Construct: Midpoint**. This is point M.

6. Use the **Circle** tool to construct the circle with center M passing through N.

7. Construct the intersection of this circle with ray AE either by selecting both and choosing **Construct: Intersection** or by using the **Point** tool to place a point where both the circle and the line change colors. Label this point T. The rest of the construction will be easier to follow if you hide the circle constructed in Step 6 by selecting it and choosing **Display: Hide Circle**.

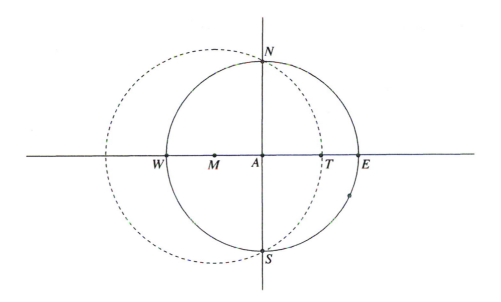

8. Use the **Segment** tool to construct line segment AT.

9. Select segment AT (it may still be selected from the construction in Step 8) and choose **Construct: Midpoint**. Call this point R.

10. Select R and segment AT and choose **Construct: Perpendicular Line**. This line is the perpendicular bisector of segment AT.

11. Construct the points where the perpendicular bisector of AT intersects the original circle centered at A. Call the intersection points U and V. U, E, and V can be vertices of a regular pentagon inscribed in the circle.

12. Construct segments VE and EU.

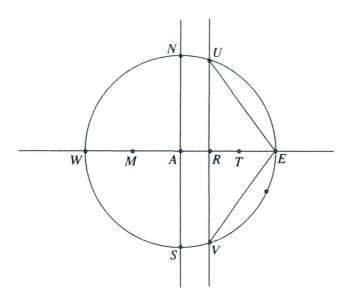

13. Construct the circles centered at V with radius VE and centered at U with radius UE. Construct the points where these circles intersect the original circle; these points are the remaining two vertices of the regular pentagon.

14. Construct the remaining sides of the pentagon to complete the figure.

43

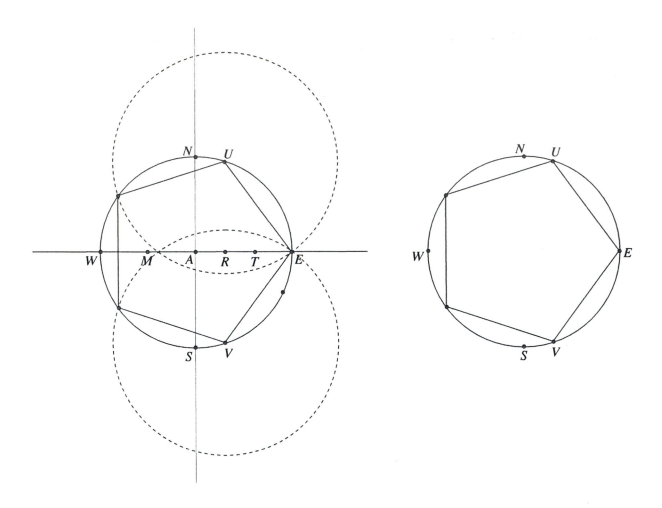

▷ **Exercise 10.** If you assume the radius AE of the original circle is one unit long, what is the exact (calculated, not measured) length of MT? In a circle of any radius, what is the ratio $\frac{MT}{AE}$?

▷ **Exercise 11.** Follow the steps in the above demonstration to construct a regular pentagon inside a circle.

9. Theoretical Origami

Companion to Chapter 3.3 of Symmetry, Shape, and Space

Since anything that is constructible by ruler and compass is also constructible using the origami postulates, we will try to model the paper folding using *The Geometer's Sketchpad*. Note that with the use of the **Transform** menu in addition to the **Construct** commands, *The Geometer's Sketchpad* is considerably more powerful than either ruler and compass or origami. We will first discuss how to do each of the origami postulates using *Sketchpad* tools. Throughout, the fold lines will be indicated by dashed lines.

More advanced users might be interested in developing a set of origami tools to replace the standard ruler and compass tools by using the **Custom Tool** option of *The Geometer's Sketchpad*.

Origami Postulate 1: *Given any two distinct points on a piece of paper, one can fold exactly one line passing through them.* [★ Origami.gsp: Postulate 1]

Origami Postulate 2: *A folded line segment can be extended at either end by an arbitrary amount (assuming the paper is big enough).* [★ Origami.gsp: Postulate 2]

Postulate 1 is easily modeled by using the **Segment** tool, while for Postulate 2 we merely replace this by the **Line** tool.

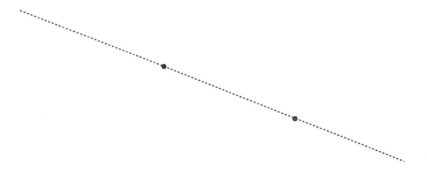

Origami Postulate 3: *Given any two distinct points on a piece of paper, one can fold the paper forming a single crease so that one point lies exactly on top of the other.* [★ Origami.gsp: Postulate 3]

By playing with a bit of paper with two marked points, one quickly figures out that this postulate involves constructing the perpendicular bisector of the line segment connecting the points.

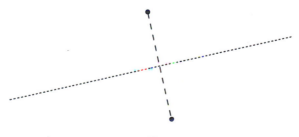

45

▷ **Exercise 1.** Model the situation and the outcome of Origami Postulate 3 using the standard tools. Hide everything except the original two points and the crease line.

Origami Postulate 4: *Given any two lines on a piece of paper, one can fold the paper forming a single crease so that one line lies exactly on top of the other.* [★ Origami.gsp: Postulate 4-1 and 4-2]

▷ **Exercise 2.** Draw two intersecting lines, and construct the line that would result from applying Origami Postulate 4. Hide everything but the original lines, the points defining these lines, and the resulting crease.

▷ **Exercise 3.** Draw two parallel lines, and construct the line that would result from applying Origami Postulate 4. Hide everything but the original lines, the points defining these lines, and the resulting crease.

Origami Postulate 5: *Given two points A and B and a line on a piece of paper with B closer to the line than it is to A, one can fold the paper forming a single crease so that point B lies on the crease while point A lies on the line.* [★ Origami.gsp: Postulate 5]

To model Postulate 5, we need to investigate the relationships a bit more. From Postulate 3, we understand that the line segment between a point and its image after a fold is perpendicular to the fold. In fact, the fold is the perpendicular bisector of the segment connecting the point and its image. We are given points A and B and a line we'll call a. The image of A (the point where it lands after the fold) is A' and A' must lie on a. The crease will go through B and will be the perpendicular bisector of AA'. Therefore, the distance from B to A will be the same as from B to A'. Consider the circle centered at B and passing through A. This will intersect the line a at two points, either one of which can be the point A'. Then the dashed crease line will be the perpendicular bisector connecting B with the midpoint of segment AA'.

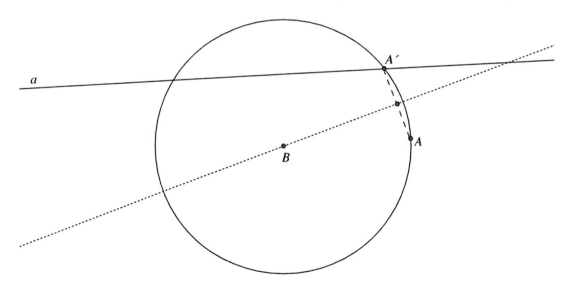

▷ **Exercise 4.** Draw two points and a line, and construct the line that would result from applying Origami Postulate 5. Hide everything but the original points and line and the resulting crease.

Recall from <u>Symmetry, Shape, and Space</u> that Origami Postulate 6 extends the traditional ruler and compass constructions. Admitting this postulate allows one to do things that are known to be impossible with ruler and compass alone, such as trisecting an arbitrary angle.

Origami Postulate 6: *Given two points A and B and two lines ℓ_1 and ℓ_2 on a piece of paper, one can fold the paper forming a single crease so that point A lies on top of line ℓ_1 while point B lies on top of line ℓ_2.* [★ Origami.gsp: Postulate 6]

We have not yet figured out how to model this postulate with *The Geometer's Sketchpad*, though theoretically this should be possible. However, it is not difficult to model an approximation.

Demonstration: Approximating Origami Postulate 6

1. Use **Construct: Line** or the **Line** tool to draw two intersecting lines. Label these lines a and b and hide the points that define these lines.

2. Use the **Point** tool to place two points A and B on the sketch.

3. Use either the point tool or the option **Construct: Point on Line** to place point A' on line a and B' on line b.

4. Construct the line segments AA' and BB' using either **Construct: Segment** or the **Segment** tool.

5. Use **Construct: Midpoint** and **Construct: Perpendicular Line** to construct the perpendicular bisectors of segments AA' and BB'.

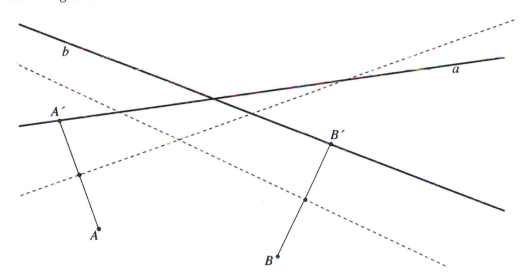

6. Note that folding across the perpendicular bisector of AA' will cause A to fall on top of A' on line a. Similarly, folding along the perpendicular bisector of BB' will make B fall on B' on line b. We want to find a single crease that will make both of these happen. Slide points A' and B' along their respective lines until the two perpendicular bisector lines coincide. It will probably take several moves before you can get them to align approximately. This is the desired crease.

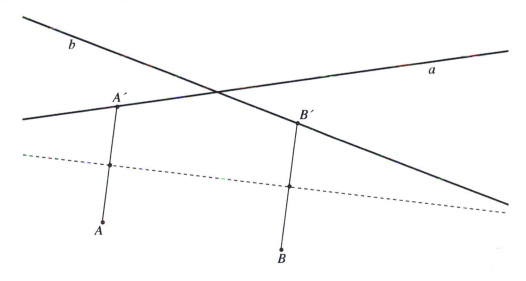

47

▷ **Exercise 5.** Follow the instructions of the preceding demonstration to approximate Origami Postulate 6 for the lines and points shown in the illustrations that accompany the demonstration.

▷ **Exercise 6.** Repeat Exercise 5 for the lines and points shown below.

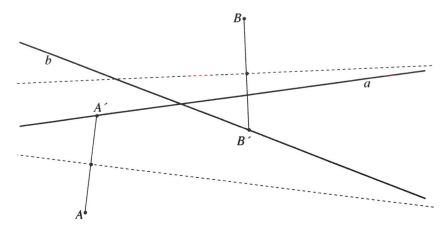

▷ **Exercise 7.** Repeat Exercise 5 for the lines and points shown below.

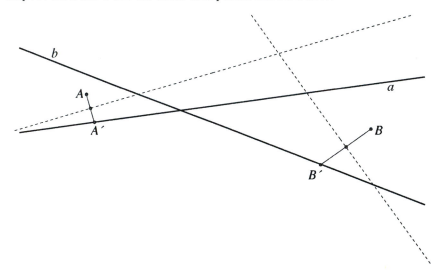

▷ **Exercise 8.** Repeat Exercise 5 for the parallel lines and points shown below.

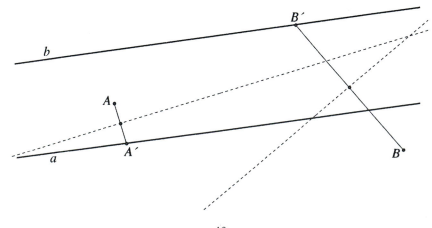

Thanks to its algebraic capabilities, *The Geometer's Sketchpad* is considerably more powerful than either ruler and compass or origami. We have already seen that we can construct polygons with an arbitrary number of sides. Furthermore, any angle is easily trisected using the **Transform** menu.

Demonstration: Trisecting an Angle [★ Trisector.gsp]

1. Construct an angle $\angle ABC$ using the **Segment** tool.
2. Select in order points A, B, and C. Choose **Measure: Angle** to measure this angle.
3. Use the **Measure: Calculate** option to compute the angle of the trisector: Click on the angle measure, then the $\div$ button, and then 3. Then click **OK**.
4. Now select point B and choose **Transform: Mark Center**.
5. Select the line segment BC and the point C, and choose **Transform: Rotate**. In the pop-up box that appears, choose the option **Marked Angle**. To fill in the **Angle** box, click on the measurement $\frac{m\angle ABC}{3}$ on the sketch to enter this quantity. Make sure that it says **About Center B**, and click on the **Rotate** button.
6. A new line segment appears, which we label BD. Measure $\angle DBC$ to verify that we have trisected the angle. Dragging point A will change the original angle, and point D will automatically move to trisect the new angle.

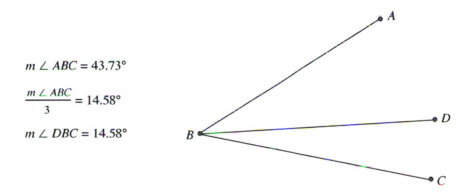

$$m \angle ABC = 43.73°$$

$$\frac{m \angle ABC}{3} = 14.58°$$

$$m \angle DBC = 14.58°$$

▷ **Exercise 9.** Construct an arbitrary angle. Trisect it by using the **Transform: Rotate** command as above.

10. Star Polygons

Companion to Chapter 3.4 of <u>Symmetry, Shape, and Space</u>

Demonstration: Star Polygons [★ Star Polygons.gsp]

The Geometer's Sketchpad is an invaluable tool for laying out the circles of dots for forming star polygons. For example, to lay out a circle of eight dots to form the star polygons $\{{8 \atop 2}\}$, $\{{8 \atop 3}\}$, etc., follow the directions below.

1. Begin by drawing a circle of a comfortable radius using either the **Circle** tool in the tool bar or the **Construct: Circle by Center + Point** menu option.

2. Select the center of the circle, and choose the menu option **Transform: Mark Center**.

3. The interior angle of an octagon is $\frac{360°}{8} = 45°$, so we want the points around the circle to be spaced at 45°. Select the point on the circle and choose **Transform: Rotate**. A small box will open up labelled **Rotate**. Make sure that the **Fixed Angle** choice is checked, and fill in the **Angle** box with **45**. Click on **Rotate** at the bottom, and another point will appear on the circle at an appropriate angle from the first.

4. Again, choose **Transform: Rotate** to get a third point, and so on until you have all eight points spaced equally around the circle.

5. Select the center of the circle and the circle itself, and use the **Display: Hide Objects** option so that all you see are the eight points spaced around the invisible circle.

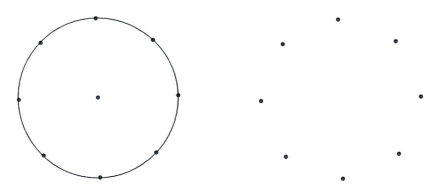

Once you have the circle of dots, it is easy to connect them (using either the **Segment** tool on the tool bar or the **Construct: Segment** menu) in order to make the various star polygons. For the compound star polygons, using a different color (use the **Display: Color** menu) for each cycle helps make the number of cycles and their lengths stand out nicely.

▷ **Exercise 1.** [SSS 3.4.4] Place 8 dots equally spaced around a circle. Construct a star polygon by connecting every second dot. Copy your circle of dots, and construct another star polygon by connecting every third dot. Repeat for every fourth dot, fifth dot, sixth dot, and seventh dot.

▷ **Exercise 2.** [SSS 3.4.5] Place 9 dots equally spaced around a circle, and draw $\{{9 \atop 1}\}$, $\{{9 \atop 2}\}$, $\{{9 \atop 3}\}$, etc.

▷ **Exercise 3.** [SSS 3.4.8] Place 12 dots equally spaced around a circle, and draw all the different $\{{12 \atop k}\}$

star polygons. For those that are compound, color each cycle a different color. You need not duplicate identical figures.

▷ **Exercise 4.** **[SSS 3.4.9]** Place 15 dots equally spaced around a circle and draw all the different $\left\{\begin{smallmatrix}15\\k\end{smallmatrix}\right\}$ star polygons. For those that are compound, color each cycle a different color. You need not duplicate identical figures.

The *Geometer's Sketchpad* can also compute the angles for the star polygons: After drawing the star polygon, choose the three points that form the angle in order and use the **Measure: Angle** menu to compute the angle.

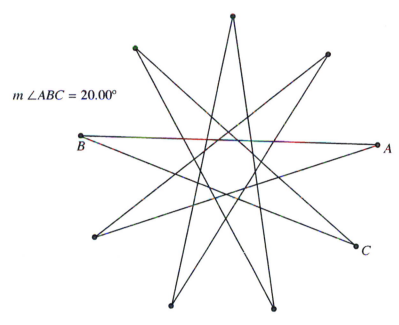

However, the exercise of doing this manually is an excellent review of basic geometric facts regarding vertical angles, supplementary angles, and the sum of the angles in a triangle or quadrilateral, so one shouldn't turn all of these computations over to the computer.

▷ **Exercise 5.** Use *The Geometer's Sketchpad* to find the vertex angles for the star polygons $\left\{\begin{smallmatrix}8\\3\end{smallmatrix}\right\}$ and $\left\{\begin{smallmatrix}9\\2\end{smallmatrix}\right\}$.

▷ **Exercise 6.** Use *The Geometer's Sketchpad* to find the vertex angles for the star polygons $\left\{\begin{smallmatrix}12\\5\end{smallmatrix}\right\}$, $\left\{\begin{smallmatrix}15\\2\end{smallmatrix}\right\}$, and $\left\{\begin{smallmatrix}15\\6\end{smallmatrix}\right\}$. Use your results to confirm the formula found in Exercise 3.4.20 of <u>Symmetry, Shape, and Space</u>.

11. Linkages

Companion to Chapter 3.5 of <u>Symmetry, Shape, and Space</u>

Demonstration: Initial Experiment [from <u>Exploring Geometry</u> by Dan Bennett]

Here is a very simple method that reproduces the results of a pantograph without actually modeling the construction: Draw a ray AB, using either the **Ray** tool or the **Construct: Ray** command. Add another point C to the ray, using either the **Point** tool or the **Construct: Point on Ray** command. Highlight both B and C, and choose the **Display: Trace Points** option. Now trace a figure with point B, and note that the trace of point C duplicates the figure at a different scale.

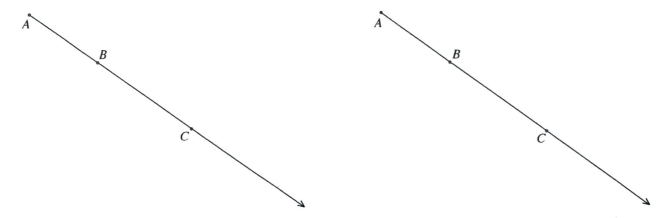

▷ **Exercise 1.** Do the experiment above, dragging point B to form any shape you like, first with point C beyond point B and then with C between A and B.

▷ **Exercise 2.** Repeat the experiment, but drag point B to form a straight line segment. Measure AB and AC (using the **Measure: Distance** command), and calculate the quotient $\frac{AB}{AC}$ using the **Measure: Calculate** command. A calculator screen will pop up and you can click on measurements already done to insert them into the formula. Place points D and E at the ends of your two line segments. Now calculate $\frac{BD}{CE}$. What do you notice?

Demonstration: The Variable-Based Triangle Linkage [★ Linkages.gsp: Variable triangle]

These instructions describe how to use *The Geometer's Sketchpad* to model the actual mechanical linkage from Exercise 3.5.1 of <u>Symmetry, Shape, and Space</u>.

1. First, we establish a standard unit measure: Draw a line segment FG about an inch long using the **Segment** tool or the **Construct: Segment** command, and use **Measure: Length** to adjust the segment until it is exactly one inch. You may need to choose **Edit: Preferences: Units** to change from centimeter measure to inches.

2. Draw a ray AH (using the **Ray** tool or the **Construct: Ray** command), and place point C on the ray (using either the **Point** tool or **Construct: Point on Ray**).

3. Draw a circle centered at A of radius FG using the **Construct: Circle by Center + Radius** command.

4. Draw another circle centered at C with radius FG, and use **Construct: Intersection** to identify point B. (If the circles do not overlap, drag point C until they do.)

5. Draw ray CB and segment AB.

6. Draw another circle centered at B that passes through point C using the **Construct: Circle by Center + Point** command.

7. Find D, the intersection of the circle of Step 6 and the ray CB.

8. Hide all of the circles and the extra intersection points.

9. Draw segment BD and use **Construct: Midpoint** to find point E.

10. Use **Display: Color** to choose three different colors for points B, D, and E.

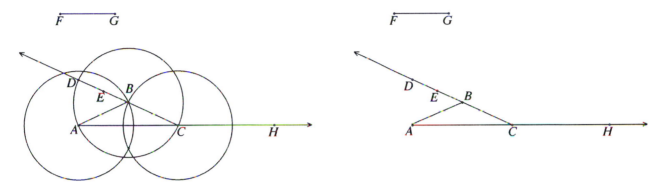

▷ **Exercise 3.** **[SSS 3.5.1]** Follow the directions above to model the variable-based triangle linkage. Select point B and choose **Display: Trace Intersection**. Drag point C along the ray AH. As C moves back and forth along the ray, describe the set of points that B traces out. Explain why this particular shape results.

▷ **Exercise 4.** **[SSS 3.5.2]** Now select point E and choose **Display: Trace Midpoint**. Drag C back and forth along the ray and describe the locus traced by the movement of E.

▷ **Exercise 5.** **[SSS 3.5.3]** Now select point D and choose **Display: Trace Intersection**. Drag C back and forth along the ray and describe the locus traced by the movement of D.

Demonstration: The Pantograph [★ Linkages.gsp: Pantograph]

1. To construct a model of a pantograph using *The Geometer's Sketchpad*, first establish a standard unit measure by drawing a line segment GH and using the **Measure: Length** command to adjust it until it is exactly one inch.

2. Construct a ray AB.

3. Draw two circles with centers at A and B, both with radius GH, using the **Construct: Circle by Center + Radius** command twice.

4. Use **Construct: Intersection** to mark point D (again, if the circles do not overlap, drag point B until they do).

5. Construct ray AD and segment BD.

6. Using either the **Point** tool or **Construct: Point on Ray**, place point E on ray AD beyond point D.

7. Select point E and line segment BD, and use the **Construct: Parallel Line** command to construct a line through E parallel to BD.

8. Use **Construct: Intersection** to place point C at the intersection of this parallel line and ray AB.

9. Draw the line through B parallel to ray AD, and place point F at the intersection of this line and line EC.

10. Now use **Display: Hide Objects** to hide both of the circles, the extra intersection point of these two circles, the rays AB and AD, and the lines EC and BF.

11. Draw line segments AD, DE, EF, FC, and BF. The result looks like a mechanical pantograph.

m GH = 1.00 in

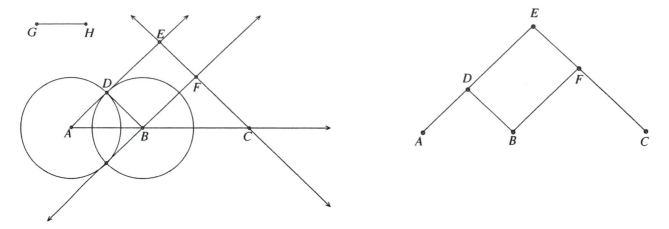

Note that this pantograph was constructed under the premise that A is fixed and B can be moved around, while preserving the parallel relationships of the lines. The same configuration works if you want to consider B as fixed and drag A around. However, if you want to move point C instead of points A or B, you will have to rebuild the simulated pantograph entirely. Therefore, Exercises 8, 9, and 11 below are for the more ambitious student. To change the scaling factor, note that AD and BD are fixed to be the length of the standard line segment GH, in this case one inch. Measure the line segment DE using the **Measure: Length** command. Moving point E until this length is also one inch will move points C and F, while keeping the configuration correct, and will give the pantograph assembly of Exercises 3.5.6 and 3.5.7 of <u>Symmetry, Shape, and Space</u>, with the point A fixed, the tracer at point B, and the pencil at point C. Choose different colors for points B and C using **Display: Color**. **Display: Trace Points** allows you to find loci. Placing two points at the extremes of each of the loci and using **Measure: Distance** allows you to approximate the scaling factors for different configurations.

▷ **Exercise 6.** [SSS 3.5.7] Select points B and C, and choose the **Display: Trace Points** command. Drag point B, and C will also move. Compare the loci of points described by B and C.

▷ **Exercise 7.** [SSS 3.5.9] Drag point E until the length of DE is 3 inches. Select points B and C, and choose **Display: Trace Points**. Drag point B, and note the difference between the loci of points described by B and C.

▷ **Exercise 8.** [SSS 3.5.8] Rebuild your *Sketchpad* pantograph so that the tracer is at C, the pencil is at B, and $DE = 1$. Draw loci at B and C, and compare them with the results of Exercise 6 above.

▷ **Exercise 9.** [SSS 3.5.10] Take your pantograph of Exercise 8 and move point E until $DE = 3$. Draw loci at B and C. What is the scaling factor?

▷ **Exercise 10.** [SSS 3.5.11] Rebuild your *Sketchpad* pantograph so that B is fixed, the tracer is at A, the pencil is at C, and $DE = 3$. Draw loci at A and C. What is the scaling factor?

▷ **Exercise 11.** [SSS 3.5.12] Rebuild your *Sketchpad* pantograph so that B is fixed, the tracer is at C, the pencil is at A, and $DE = 3$. Draw loci at A and C. What is the scaling factor?

54

▷ **Exercise 12.** **[SSS 3.5.13]** Use *The Geometer's Sketchpad* to model the linkage below in which point A is fixed, point B is used to trace a figure, and there is a pencil at point C. Find the scaling factor.

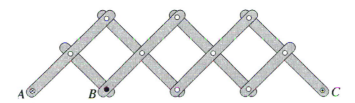

▷ **Exercise 13.** **[SSS 3.5.15]** Use *The Geometer's Sketchpad* to model the linkage below in which point A is fixed, point B is used to trace a figure, and there is a pencil at point C. Find the scaling factor.

▷ **Exercise 14.** **[SSS 3.5.16]** Use *The Geometer's Sketchpad* to model the linkage below in which point A is fixed, point B is used to trace a figure, and there is a pencil at point C. Find the scaling factor.

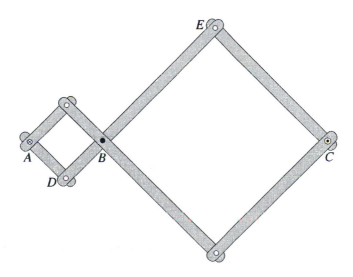

▷ **Exercise 15.** **[SSS 3.5.17]** Use *The Geometer's Sketchpad* to model the linkage below in which point A is fixed, point B is used to trace a figure, and there is a pencil at point C. Find the scaling factor.

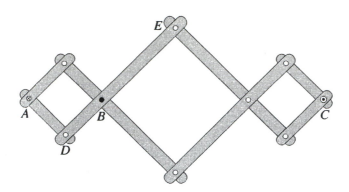

▷ **Exercise 16.** **[SSS 3.5.18]** Design three different parallelogram linkages that act as pantographs with scaling factor 3.

▷ **Exercise 17.** **[SSS 3.5.19]** Model Peaucellier's linkage using *The Geometer's Sketchpad*. Points X and Y are fixed. Use circles to ensure that $XY = YD$, $XA = XC$, and $AB = BC = CD = DA$. Describe the locus of points traced by the movement of point B.

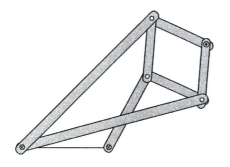

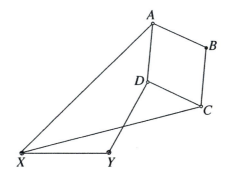

12. Regular and Semiregular Tilings

Companion to Chapter 4.1 of Symmetry, Shape, and Space

▷ **Exercise 1.** Two of the three regular tilings, squares and equilateral triangles, can be formed by a grid of straight lines. Use the **Construct: Line** command or the **Line** tool, **Transform: Translate**, and **Transform: Rotate** to generate these patterns.

▷ **Exercise 2.** Create a regular hexagon using either Method 2 or 3 from Section 3. Copy and paste the interior of the hexagon. (You may need to hide some key points.) Copy and paste hexagons to form a tiling. You should not need to rotate any of them. How many colors should you use for the interiors of the hexagons so that no two tiles that share a border are the same color?

Demonstration: Regular Polygon Tiles [★ Polygonal Tiles.gsp]

For this section, Method 3 of Section 3 for constructing regular polygons is preferred, since this method gives polygons with a standard edge length which will then fit together nicely. You can choose to use *The Geometer's Sketchpad* to find the vertex angles of the polygons rather than using the formula $\frac{(n-2)180°}{n}$. Choose three vertex points in order, and choose **Measure: Angle**.

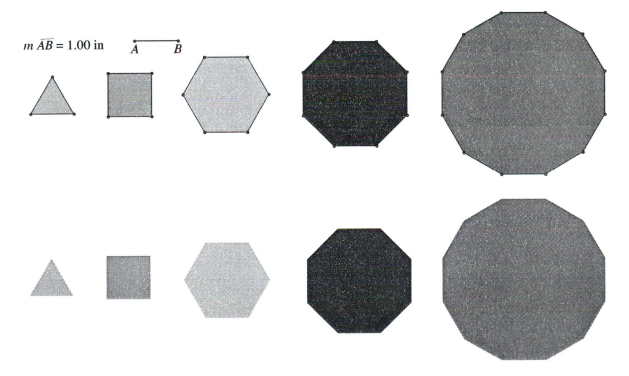

$m\,\overline{AB} = 1.00$ in

1. Construct regular polygons with 3, 4, 6, 8, and 12 sides using Constructing Regular Polygons: Method 3 from Section 3.

2. Select all of the vertices of each polygon in order (clockwise or counterclockwise), and use the **Construct: Interior** and **Display: Color** commands to create and color each polygon.

3. Copy and paste the polygon interiors. (You may need to hide some key points.) You should be able to rotate the pasted polygons freely without resizing.

▷ **Exercise 3.** Follow the general instructions above to form a family of basic regular polygons: equilateral triangle, square, hexagon, octagon, and dodecagon. Copy and paste their interiors.

Demonstration: Tiling Method 1 [★ Tiling 4.8.8.gsp]

To create a tiling, for example 4.8.8, either use **File: New Sketch** to open a new drawing page, or use **File: Document Options: Add Page** to add a blank page to your sheet of polygons from the previous demonstration.

1. Copy the square and octagon interiors from your polygon page, and paste them on your blank page. Now you have to fit them together. The **Selection Arrow** tool from the tool bar is used to drag the polygons around without resizing or changing orientation. To rotate the polygons in order to join them to form tilings, use the **Rotate Selection Arrow** tool from the tool bar: Click on the **Selection Arrow** and slide to the right as you hold down the mouse button to choose the **Rotate Selection** tool.

2. Click on the square and rotate it about 45° until it fits onto the diagonal octagon edges. You may need to adjust it several times before getting it to fit nicely.

3. You will need to paste additional copies of the square and octagon and use the **Selection Arrow** tool to drag them into position. Note that in order to color tiles that share an edge with different colors (so you can tell where one stops and the other starts), you must use two different colors for the octagons.

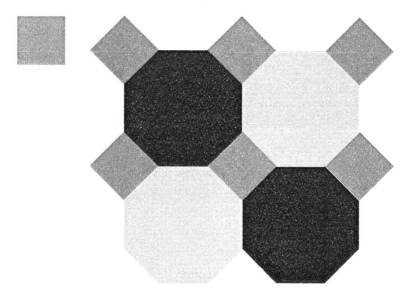

▷ **Exercise 4.** Use Tiling Method 1 to draw a section of the tiling 3.12.12. How many colors do you need to use for the polygons so that no two that share a border are the same color?

▷ **Exercise 5.** Use Tiling Method 1 to draw a section of the tiling 3.4.6.4. How many colors do you need to use for the polygons so that no two that share a border are the same color?

▷ **Exercise 6.** Use Tiling Method 1 to draw a section of the tiling 3.3.4.3.4. How many colors do you need to use for the polygons so that no two that share a border are the same color?

▷ **Exercise 7.** Use Tiling Method 1 to draw a section of the tiling 3.3.3.3.6. How many colors do you need to use for the polygons so that no two that share a border are the same color?

Instead of placing each tile individually, it is easier to join them in blocks or groups. For example, at some home improvement stores, you can find concrete blocks that look like this, with a groove separating the octagon from the square:

These tiles can be joined together to form the tiling 4.8.8, but with the convenience of only having to deal with one shape of tile. Such building blocks are called *fundamental regions* for the tiling, and the tiling consists of repeats of the fundamental region, without rotation or reflection. Of course, a tiling has many possible fundamental regions. Here are some others for the pattern 4.8.8.

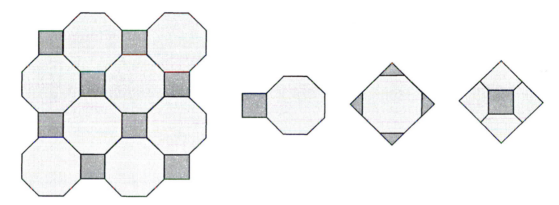

Finding a fundamental region for a tiling makes drawing the tiling with a computer easier: You just need to cut and paste the region and slide it around, without having to rotate tiles each time.

▷ **Exercise 8.**   Find two different fundamental regions for the tiling 3.12.12.

▷ **Exercise 9.**   Find two different fundamental regions for the tiling 3.4.6.4.

▷ **Exercise 10.**   Find two different fundamental regions for the tiling 3.3.4.3.4.

▷ **Exercise 11.**   Find two different fundamental regions for the tiling 3.3.3.3.6.

While rotating polygons with the **Rotate Selection Arrow** tool and sliding them around with the **Selection Arrow** tool duplicates the natural process that one would use with actual tiles, it is almost impossible to get the edges aligned precisely with this technique. For presentation quality sketches, another method should be used.

Demonstration: Tiling Method 2

For this method, you will still need your sketch containing the family of basic polygons. To draw the tiling 4.8.8, for example, copy and paste the octagon, including its vertices and edges, to a new page. We label this octagon $ABCDEFGH$ for the purposes of this explanation.

1. Select the vertex A and click on **Transform: Mark Center**.

2. Select the edge AB and the vertex B, and choose **Transform: Rotate** to rotate the segment by 90° (or −90° if the rotation places the new edge in the interior of the octagon). This forms a new edge AB'.

3. Rotate the new line segment AB' around its free endpoint B' by 90° to get edge $B'A'$. Construct $A'B$ to complete a square that shares one edge with the octagon.

4. Select the vertices of the square in order, and use **Construct: Quadrilateral Interior** and **Display: Color** to form and color the interior of the square.

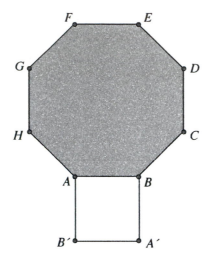

5. Rather than repeating this process for the other three squares around the octagon, select first one vertex of the square and then the vertex of the octagon where you want to place the corresponding vertex of the new square, and then choose **Transform: Mark Vector**.

6. Select the square, its vertices, and its edges and choose **Transform: Translate**. This should give a copy of the square in the new position.

7. Repeat to get the other squares and octagons that form the tiling. For a tiling where the repeats of a particular polygon are not simply translates of the first one drawn, you may have to use **Transform: Reflect** to reflect across the appropriate edge or **Transform: Rotate** to rotate by an angle you have computed.

▷ **Exercise 12.** Use Tiling Method 2 to draw a section of the tiling 4.6.12. How many colors do you need to use for the polygons so that no two that share a border are the same color?

▷ **Exercise 13.** Use Tiling Method 2 to draw a section of the tiling 3.6.3.6. How many colors do you need to use for the polygons so that no two that share a border are the same color?

▷ **Exercise 14.** Use Tiling Method 2 to draw a section of the tiling 3.3.3.4.4. How many colors do you need to use for the polygons so that no two that share a border are the same color?

▷ **Exercise 15.** Find two different fundamental regions for the tiling 4.6.12.

▷ **Exercise 16.** Find two different fundamental regions for the tiling 3.6.3.6.

▷ **Exercise 17.** Find two different fundamental regions for the tiling 3.3.3.4.4.

13. Irregular Tilings

Companion to Chapter 4.2 of Symmetry, Shape, and Space

The two tiling methods of the previous section adapt well to irregular tilings. The first method allows you to play with a tiling to see how the pieces fit together. The second method produces better pictures, but you have to know what pieces to copy and where they go in order to use it. Tiling Method 1 is better for experimenting to figure out how to fit the pieces together. Then you can translate and rotate the pieces using the **Transform** menu for precise drawings using Tiling Method 2.

▷ **Exercise 1.** **[SSS 4.2.1]** Construct an arbitrary triangle using the **Segment** tool. Construct and copy the interior. Fit copies together (using Tiling Method 1) to tile the plane.

Demonstration: Tiling with an Arbitrary Triangle [★ Triangular tiling.gsp]

We will do Exercise 1 using Tiling Method 2.

1. Construct an arbitrary triangle $\triangle ABC$ using the **Segment** tool. Select the vertices and use **Construct: Triangle Interior** to form the interior.
2. Construct the midpoint of side BC of your triangle. Double-click on the midpoint to mark it as the rotation center (or use **Transform: Mark Center**).
3. Choose **Select All** from the **Edit** menu. Then choose **Transform: Rotate**, and rotate by 180°.
4. Now select the endpoints (A and B) of a different side of the original triangle, and choose **Transform: Mark Vector**.
5. Choose **Edit: Select All** and **Transform: Translate**.
6. Now select the endpoints (A and C) of the third side of the original triangle, and choose **Transform: Mark Vector**.
7. Choose **Edit: Select All** and **Transform: Translate**.

By now, you should have eight copies of your original triangle fitting together to tile a section of the plane. Choosing translation vectors carefully will let you tile as large a section as you want. You can also translate more than once in the direction of a vector to make more copies. My personal preference is moving bigger groups along longer vectors. We translated the block of eight copies along AD and AE for our final picture below.

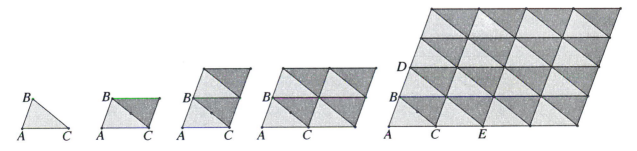

Note that there are other ways to create the tiling. You could have done rotations instead of translations at many points. We shaded the tiles with two colors to make the picture easier to decipher.

61

▷ **Exercise 2.** Construct an arbitrary trapezoid using the **Segment** tool. Fit copies together to tile the plane.

▷ **Exercise 3.** [SSS 4.2.3] Draw the tiling generated by the nonconvex quadrilateral below. Your tiling should contain at least 12 repeats of the quadrilateral.

In Exercise 3 above, if you use Tiling Method 2, you only need to rotate the tile once. The rest of the tiling can be generated by translating the pair of tiles along carefully chosen vectors. Translating more than the first pair of tiles can generate the drawing faster, but you have to be a little more careful choosing your translation vectors. We rotated the original quadrilateral by 180° around the midpoint of CD. Then we translated the resulting figure along AC to get four copies of the original tile. Finally, we translated along BD to get eight copies of the tile.

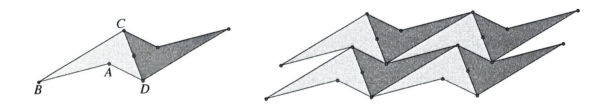

▷ **Exercise 4.** Use the **Transform: Rotate** command and the **Segment** tool to construct the trapezoidal reptile from Exercise 4.2.4 of <u>Symmetry, Shape, and Space</u> shown below. Then use the **Transform: Reflect** command to create the larger copy using four copies of the original trapezoid. Note that double-clicking on an edge will mark it as the mirror. Holding the mouse button and dragging will allow you to select a group of objects. If you select too many, clicking on the ones you don't want will deselect them. Repeat the process for an arbitrary trapezoid, showing that in general trapezoids are not reptiles.

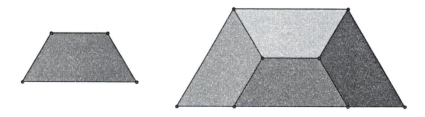

▷ **Exercise 5.** [SSS 4.2.8] Show that each of the following is a reptile. Note the proportions given by the grid. Use the **Rotate**, **Reflect**, and/or **Translate** commands under the **Transform** menu to combine four copies of each tile to create the larger version.

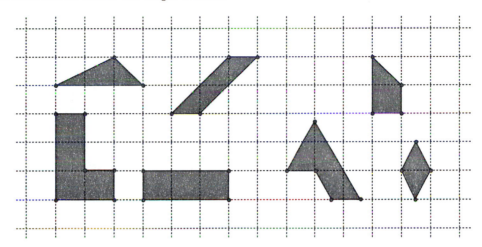

Creating Escher-Style Tilings

Many of M.C. Escher's tiling pictures were created using translation, glide reflection, or rotations on a polygonal tile. We will next investigate how to create such figures using *The Geometer's Sketchpad*, which does the required transformations very nicely. Many of these irregular tilings require cutting along one edge of a shape and translating or rotating the cut-off piece to another edge. *The Geometer's Sketchpad* cannot cut random curves like one can with scissors and cardboard, but jagged cuts using line segments can still demonstrate the properties of a tiling. We will investigate each type of transformation in turn.

Translation

Recall that squares, rectangles, parallelograms and hexagons have pairs of parallel sides and so form grids that allow us to use parallel translation. To create a tiling using this operation, follow the process described in Exercise 6 below and refer to the illustration on the next page. You can drag the vertices A, B, C, or the vertices along the jagged path from A to B to modify the tiling.

Demonstration: Parallelogram Tiling Using Translation [★ Tiling by translation.gsp]

1. Construct a line segment AB and a point C above the segment.
2. Select A and B in order and choose **Transform: Mark Vector**.
3. Select C and **Transform: Translate** to get point C'.
4. Constructing line segments AC, CC', and $C'B$ finishes your parallelogram. (You can also use the **Construct: Parallel Line** and **Construct: Intersection** commands to create a parallelogram.)
5. Sketch a string of several short line segments to form a jagged path from A to B.
6. Select first A and then C to mark AC as the translation vector using **Transform: Mark Vector**.
7. Select the segments forming the jagged path and their vertices, and choose **Transform: Translate** to move a copy to edge CC'.
8. Hide the original segments AB and CC'.
9. Create a column of tiles using your new tile by choosing **Edit: Select All** and **Transform: Translate** several times (still in the AC direction).

10. Then choose *AB* as the transformation vector, select the whole column of tiles, and translate it to cover the plane (or at least your drawing window). Your final drawing should contain at least 20 copies of your original tile.

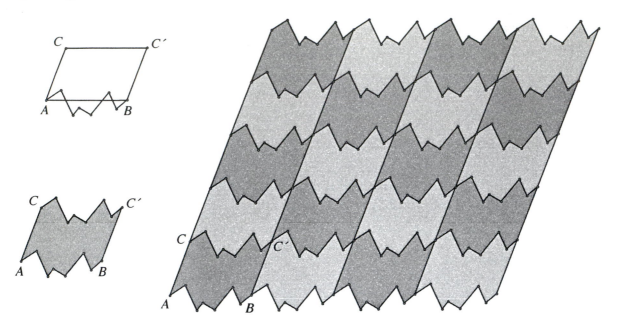

▷ **Exercise 6.** [SSS 4.2.9] Create an irregular tiling from a parallelogram, using translation of both pairs of parallel sides.

▷ **Exercise 7.** [SSS 4.2.10] Create an irregular tiling from a hexagon, using translation of all three pairs of parallel sides.

Glide Reflection

Recall that squares, rectangles, parallelograms, and hexagons have pairs of parallel sides and so form grids that allow us to use glide reflection. Glide reflections require a reflection and a translation. On a parallelogram, reflect across a line perpendicular to and through the endpoint of the side of the figure. Then translate along one of the diagonals of the parallelogram. See the following sequence of steps and the figure at the top of the next page.

Demonstration: Parallelogram Tiling Using Glide Reflection [★ Tiling by glide reflection.gsp: Creating the tile]

1. Construct a parallelogram *ABDC*.
2. Select line segment *AB* and point *B*. Then choose **Construct: Perpendicular Line** to construct the line perpendicular to side *AB* and through *B*.
3. Create a jagged path along the side *AB* using line segments.
4. Double-click on the perpendicular line to make it the mirror line, or select it and choose **Transform: Mark Mirror**.
5. Select the jagged path and reflect it across the perpendicular line using **Transform: Reflect**.
6. Select first point *B* and then point *C*, and choose **Transform: Mark Vector** to make this the translation vector.

64

7. Select the reflected path and choose **Transform: Translate**.

8. Hide the original reflected path and the perpendicular line.

9. Hide the straight line segments AB and CD. You now have a single tile. Select the vertices of your tile in order and use **Construct: Polygon Interior** and **Display: Color** to form and color your tile.

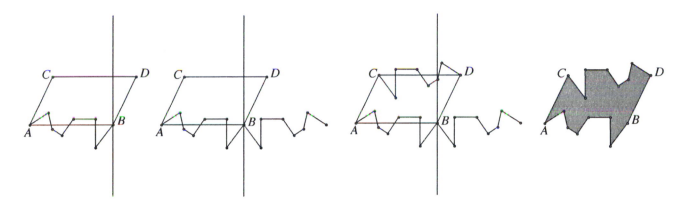

Careful selecting, copying, reflecting, and translating will allow you to tile the plane with your irregular tile. For the picture below, we copied the original tile, undid the hide command using **Display: Show All Hidden**, deselected the perpendicular line as the reflection line, and redid the **Display: Hide Objects** command. Then we reflected the copy of the tile and hid the copy, leaving only the original and the new reflected tile. [★ Tiling by glide reflection.gsp: Fitting tiles by hand] Next, we made several copies of each orientation of the tile, being careful to spread them far enough apart that we were able to select a whole tile (hold the mouse button and slide it to select multiple objects) without selecting parts of another. Then we put the tiles together by dragging them into position to get the approximate tiling below. This is basically the approach used in Tiling Method 1 of Section 12. Fitting the tiles together precisely is quite difficult. There must be an easier way.

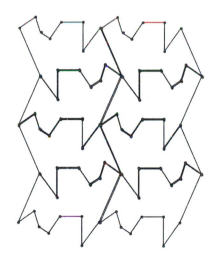

The next picture was created using a modification of Tiling Method 2 of Section 12. As above, reflect your original tile across the perpendicular line, which you have unhidden for this purpose. Mark the translation vector BC, and translate the reflected tile into position along CD. Now hide the first reflection of the tile. You have a pair of tiles stacked on top of each other, one in each configuration needed. Translate this double tile in both directions to fill the drawing window. As with the translation tiling, you can drag the vertices A, B, C, or the vertices along the jagged path from A to B to modify the tiling. [★ Tiling by glide reflection.gsp: The tiling]

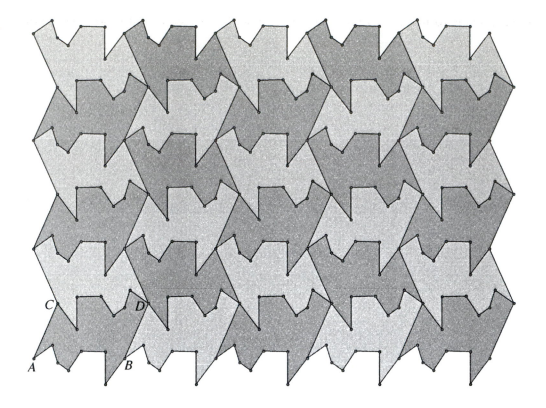

Note that the following exercise uses a rectangular grid instead of a parallelogram grid.

▷ **Exercise 8.** [SSS 4.2.11] Create an irregular tiling from a rectangle, using glide reflection of both pairs of parallel sides.

Midpoint Rotation

One can use a basic grid formed by any quadrilateral or triangle for midpoint rotations, because we will not need parallel sides. We will demonstrate the technique using parallelograms as in Symmetry, Shape, and Space. The exercises duplicate those in the primary text.

Demonstration: Parallelogram Tiling Using Midpoint Rotation [★ Tiling by midpoint rotation.gsp: Creating the tile]

1. Construct a parallelogram $ABCD$.
2. Construct the midpoint of each side using **Construct: Midpoint**.
3. Form a jagged path from A to E, the midpoint of AB.

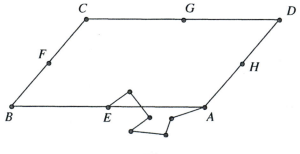

4. Double-click on E to mark it as the center of rotation (or select E and choose **Transform: Mark Center**).

5. Select your jagged path from A to E, and choose **Transform: Rotate**. Type **180** in the box and click the **Rotate** button.

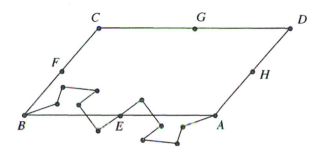

6. Repeat the process in Steps 3–5 on the other sides of the parallelogram.

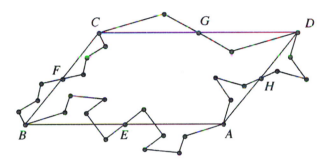

7. Hide the original parallelogram, but leave the midpoints visible. Select the vertices in order around the tile, and create and color the interior. For a complicated tile, it may help to color-code the vertices: one color for the vertices of the original parallelogram, another for the midpoints, and a third for the other vertices along the jagged paths.

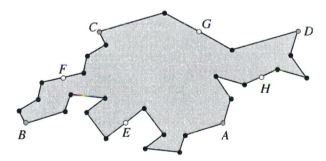

This is an irregular tile generated by midpoint rotation on each side of a parallelogram. Next we will tile the plane. First we will fit together four copies of the new tile. [★ Tiling by midpoint rotation.gsp: Fitting 4 tiles together]

8. Mark E as your center of rotation. Choose **Edit: Select All** and then **Transform: Rotate** by 180°. This should give you two copies of the tile fitting together along edge AB.

9. Leaving the new tile highlighted, double-click on F' to make it the new center of rotation. Choose **Transform: Rotate** (still by 180°) to fit a third copy of the tile around point A.

10. Leaving the new, third tile highlighted, double-click on G'' and choose **Transform: Rotate**. This should give you four copies of the tile around A and a fundamental region for the tiling.

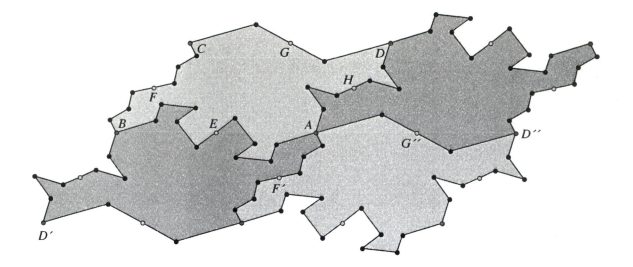

11. Select C and D' and choose **Transform: Mark Vector**. Then choose **Edit: Select All** followed by **Transform: Translate** to copy the fundamental region into a column of blocks of four tiles.

12. Mark BD'' or a similar vector as the translation vector, and translate the column generated in Step 11 to tile the plane.

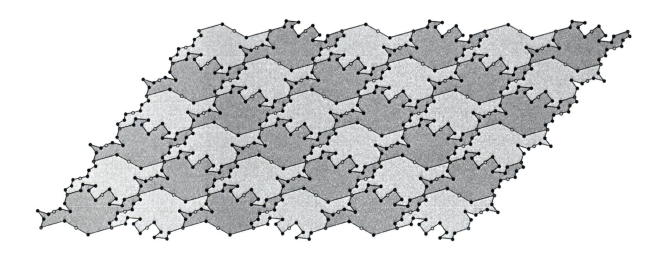

▷ **Exercise 9.** [SSS 4.2.12] Create an irregular tiling from an equilateral triangle grid, using midpoint rotation on each side.

▷ **Exercise 10.** [SSS 4.2.13] Create an irregular tiling from a parallelogram grid, using a translation on one pair of parallel sides and a midpoint rotation on each of the other sides.

Side Rotation

Side rotation is very similar to midpoint rotation, but the angle of rotation will change depending on the grid you use and the direction you want to rotate. The jagged piece cut from one side of a square can be rotated to either adjacent side by rotating through an angle of 90° or −90°. Once the tile is created, fit copies of the tile into a fundamental region and tile the plane as in the demonstration for midpoint rotation.

Demonstration: Square Tiling Using Side Rotation [★ Tiling by side rotation.gsp: Creating the tile]

1. Construct a square $ABCD$. (See Section 0 if you need help with this.)

2. Form a jagged path from A to B.

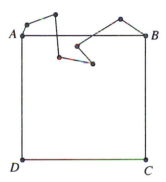

3. Mark B as the center of rotation, select the jagged path, and rotate it by 90° to get the tile shown below on the left. Alternatively, one could mark A as the center of rotation, select the path, and rotate it by −90° to get the tile on the right. Either tile will tile the plane. (How can you tile the plane using some of each tile?) We will work with the first one.

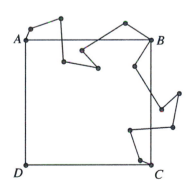 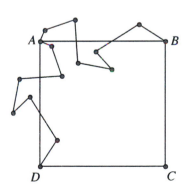

5. Hide line segments AB and BC. At this point, one could go on and perform a similar operation on the other pair of adjacent sides, but you can do that on your own. Construct and color the interior of your tile.

6. Mark B as the center of rotation. Choose **Edit: Select All**. Then choose **Transform: Rotate**, put **90** in the box, and click **Rotate** to fit two tiles together. [★ Tiling by side rotation.gsp: Fitting 4 tiles together]

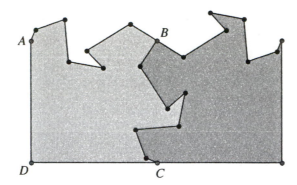

7. Rotate the new tile twice more to form a fundamental region consisting of four tiles.

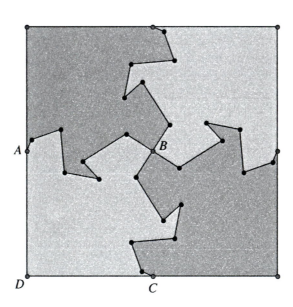

8. Select the entire region and translate it along the edges to tile the plane. [★ Tiling by side rotation.gsp: The tiling]

70

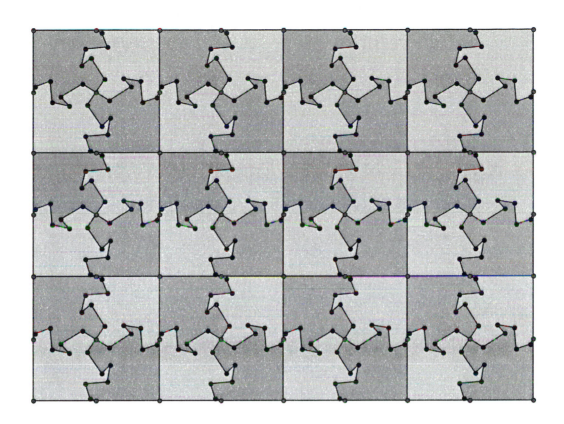

▷ **Exercise 11.** **[SSS 4.2.14]** Create an irregular tiling from an equilateral triangle grid, using side rotation on one pair of sides and midpoint rotation on the other side.

14. Penrose Tilings

Companion to Chapter 4.3 of <u>Symmetry, Shape, and Space</u>

The Geometer's Sketchpad can easily generate pictures of the Penrose tiles. It can rotate the tiles to begin some of the tilings. However, there is usually too much trial and error involved in expanding a tiling to make the software helpful. We suggest creating the tiles with *The Geometer's Sketchpad*, printing these templates for the tiles onto cardstock or stiff paper, cutting them out, and working by hand.

The tiles involve copies of the golden triangle created in Section 8: The Pentagon and the Golden Ratio. Hence, they can be constructed using only the traditional ruler and compass. The **Transform** menu options of *The Geometer's Sketchpad* allow much easier construction, which is the method we follow here.

The angles for the kite and dart are computed in Exercise 4.3.2 of <u>Symmetry, Shape, and Space</u>. Using the **Transform: Rotate** command and these angles, we will generate the shapes of the Penrose tiles.

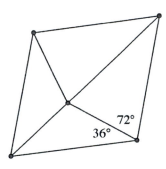

Demonstration: Creating Penrose Tiles

1. Construct a line segment AB of length 1, using **Measure: Length** to verify the length.

2. Rotate the segment by 108° twice around endpoint A to form segments AB' and AB''.

3. Construct the line segments BB' and $B'B''$.

4. Hide the line segment AB'. Select the vertices in order and construct and color the interior of the dart tile. [★ Penrose Tiles.gsp: Dart]

$AB = 1.00$ in

$m\angle BAB' = 108.00°$

$m\angle B'AB'' = 108.00°$

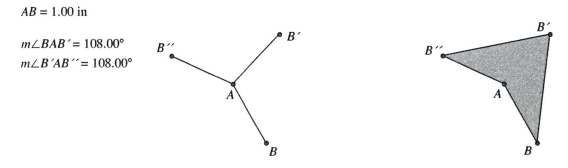

5. Now copy your first three lines, AB, AB', and AB'', to a new page to make sure you have the same lengths for the kite.

6. Construct the line (not line segment) AB'.

7. Draw the circle centered at B with radius BB', using either the **Circle** tool or **Construct: Circle by Center + Point**.

8. Use **Construct: Intersection** with the circle and line AB' to locate point C.

9. Construct line segments BC and $B''C$ to form the kite.

10. Hide the circle, line and line segment AB', and point B'. Select the vertices in order, and construct and color the interior of the kite tile. [★ Penrose Tiles.gsp: Kite]

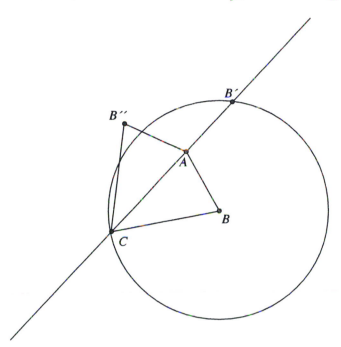
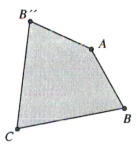

▷ **Exercise 1.** Construct a kite and a dart as above.

▷ **Exercise 2.** Using the **Transform: Rotate** command, copy five darts around the vertex B' to form the Star.

▷ **Exercise 3.** Using the **Transform: Rotate** command, copy five kites around the tail C to form the Sun.

▷ **Exercise 4.** Using the **Transform: Translate** command, cover a page with kite-dart rhombi, print several copies, cut out the pieces, and use them to create a Penrose tiling.

15. Kaleidoscopes

Companion to Chapter 5.1 of <u>Symmetry, Shape, and Space</u>

We suggest that you first check out the cool sample sketch that came with *The Geometer's Sketchpad 4.0* in the folder **Sketchpad: Samples: Sketches: Fun: Half head.gsp** and play with it.

Demonstration: Modeling a Two-Mirror Kaleidoscope [★ Kaleidoscope.gsp]

1. To build a theoretical two-mirror kaleidoscope, start by constructing an angle formed by two roughly equal line segments *AB* and *AC*.

2. Inside this angle, draw an irregular asymmetric polygon.

3. Measure the angle ∠*BAC* by choosing the three points in order and using **Measure: Angle**.

4. Select the segment *AB* and choose **Transform: Mark Mirror**. The line segment should flash.

5. Now select the polygon and the segment *AC* and use the **Transform: Reflect** option.

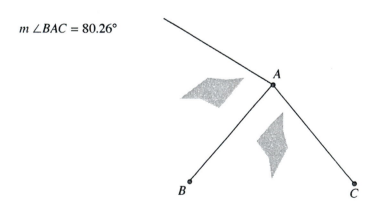

6. Now repeat the process, reflecting the polygon and the segment *AB* across the line *AC*. Reflect across the reflected line segments until you have a circular pattern of reflections (as shown in the following illustration). Continue to reflect the line segments and the polygon until you have ten or twelve copies.

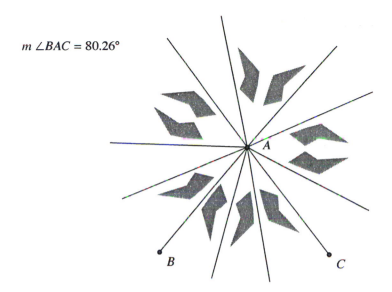

$m \angle BAC = 80.26°$

Now drag the point C around to change the angle $\angle BAC$. You will notice that at 90° (or as close as you can get), the images overlap completely.

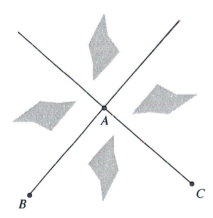

$m \angle BAC = 89.88°$

At 72°, the line segments fit nicely around a circle, except that each polygon image is overlapped by its reflection (shaded in two colors below to make this more obvious).

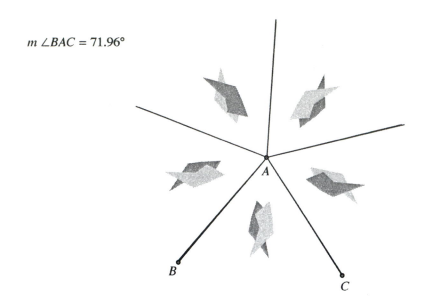

$m \angle BAC = 71.96°$

▷ **Exercise 1.** **[SSS 5.1.16]** Which angles give you a whole number of images? Are there an equal number of right-handed and left-handed images? How many images are formed? Give general formulae for the angles that work and the relationship between the angle and the number of polygons.

▷ **Exercise 2.** **[SSS 5.1.18]** Repeat the process above but place a symmetric figure inside of the angle. Which angles give you a whole number of images? How many images are formed? Give general formulae for the angles that work and the relationship between the angle and the number of images.

The experiments above should lead you to the realization that all kaleidoscope angles are of the form $\frac{360°}{2n}$, where $n \geq 2$ is the number of right-handed images (which must equal the number of left-handed images).

▷ **Exercise 3.** Draw a circle and two radial lines AB and AC. Use the instructions of the demonstration above to model a two-mirror kaleidoscope with central angle 60°.

▷ **Exercise 4.** Draw a circle and two radial lines AB and AC. Use the instructions above to model a two-mirror kaleidoscope with central angle 45°.

▷ **Exercise 5.** Draw a circle and two radial lines AB and AC. Use the instructions above to model a two-mirror kaleidoscope with central angle 30°.

Demonstration: Modeling a Three-Mirror Kaleidoscope

To model a three-mirror kaleidoscope, you must first construct one of the three Coxeter triangles: the equilateral 60°-60°-60° triangle, the isosceles right triangle with angles 45°-45°-90°, or the 30°-60°-90° triangle. We will model the 60°-60°-60° kaleidoscope and leave the others for you to do.

1. Construct an equilateral triangle (either using the classic construction by circles or by rotating a line segment by 60° twice).

2. Draw an asymmetric object and place it within the triangle.

3. Select one of the sides of the triangle and choose **Transform: Mark Mirror**.

4. Select the rest of the triangle and your object, and choose **Transform: Reflect**.

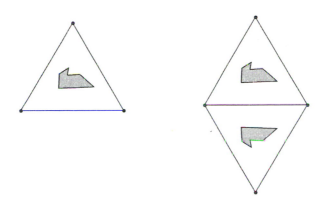

5. Repeat the process, reflecting through the other sides of the original triangle and then reflecting across the sides of the reflected triangles, and so on.

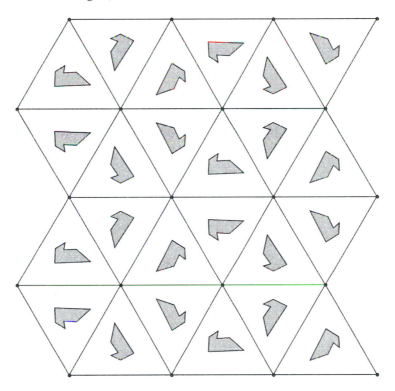

▷ **Exercise 6.** Model a kaleidoscope built using a mirror assembly forming an isosceles right triangle with angles 45°-45°-90°.

▷ **Exercise 7.** Model a kaleidoscope built using a mirror assembly forming a right triangle with angles 30°-60°-90°.

16. Rosette Groups

Companion to Chapter 5.2 of <u>Symmetry, Shape, and Space</u>

Creating rosette patterns, which have a central point of symmetry, is a lot like modeling two-mirror kaleidoscopes, but with the added interest of allowing rotations about the center point in addition to reflections. Recall the notation introduced in <u>Symmetry, Shape, and Space</u>: C_n denotes the symmetry type of an object with an n-fold rotation (through $\frac{360°}{n}$) and no reflections, and D_n denotes an object with a n-fold rotation and a reflection (and therefore, n reflections).

Demonstration: Drawing Rosette Patterns C_6 and D_6 [★ Rosette Groups.gsp]

1. We will create an object with symmetry group C_6 by first placing a central point A on the sketch.

2. Now draw a asymmetric object, made out of line segments and polygons. This may include the central point or not. Below are two possible scenarios.

3. In order to have symmetry group C_6, the object must have a $\frac{360°}{6} = 60°$ rotation. Select the point A and choose **Transform: Mark Center**.

4. Select the object and choose **Transform: Rotate**. A box labeled **Rotate** will appear. Make sure that the **Fixed Angle** option is checked and that it says **About Center A** at the bottom. Fill in the **Angle** blank with $60°$. Choose the **Rotate** button at the bottom, and another copy of the object will appear at the correct angle.

5. Click on **Transform: Rotate** again (using the same angle), and a third copy will appear.

6. Repeat until you have six copies arranged in a circle around A. (Note that one more rotation would give a seventh copy precisely on top of the first.) The resulting figure has rosette group C_6.

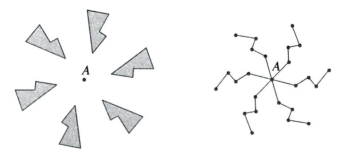

7. There are two ways to draw an object with symmetry group D_6. In the first method, begin with an object with symmetry type C_6. Draw a line segment from A out though the C_6 picture. Select this line segment and choose **Transform: Mark Mirror**. Then select the entire C_6 pattern and choose **Transform: Reflect**. This will result in a D_6 pattern.

8. Another method is to start with a central point A and a single asymmetric motif. Draw a line segment from A, and select this for **Transform: Mark Mirror**. Select your motif and use **Transform: Reflect** to get a mirror image. Now choose A to be the center of rotation, using **Transform: Mark Center**. Select your original motif and its mirror image and use **Transform: Rotate** to rotate both figures by 60° about the center A. Repeat to get six double copies spaced around in a circle. In both methods, after you're done hide the line of reflection and its far endpoint using **Display: Hide Objects**.

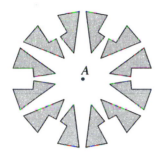 

▷ **Exercise 1.** [SSS 5.2.10] Draw a figure that has 8-fold rotational symmetry but no lines of reflection, and thus rosette group C_8.

▷ **Exercise 2.** [SSS 5.2.11] Draw a figure that has reflectional symmetry and an 8-fold rotation. This will have rosette group D_8. How many lines of reflection does your figure have?

▷ **Exercise 3.** [SSS 5.2.12] Draw a figure that has a line of reflection, but no rotational symmetry, and so has rosette group D_1.

▷ **Exercise 4.** [SSS 5.2.13] Draw a figure that has rosette group C_1.

▷ **Exercise 5.** [SSS 5.2.14] Draw a figure that has rosette group C_2.

▷ **Exercise 6.** [SSS 5.2.15] Draw a figure that has rosette group D_2.

▷ **Exercise 7.** [SSS 5.2.18] Draw another figure that has the same symmetries, and so the same rosette group D_4, as the square.

In addition to drawing figures of various symmetry types, *The Geometer's Sketchpad* can help users in the analysis of the rosette groups. When discovering the different configurations of a figure, most people find that using color as a visual aid is extremely helpful. Many also find that actually cutting out a copy of the figure and moving it by hand helps keep things straight. *The Geometer's Sketchpad* allows one to easily color, copy, rotate, and reflect a figure to determine which operations give a new configuration and which are equivalent to a previous operation. We will follow the development in <u>Symmetry, Shape, and Space</u> by leading you through the analysis of an equilateral triangle and having you create similar figures for the square.

Demonstration: Symmetries of an Equilateral Triangle [★ Symmetries of a triangle.gsp]

We will create a series of pictures representing all distinct configurations of an equilateral triangle where the transformed triangle is in the same orientation as the original. Since this text is printed in black and white, we use solid, dashed, and dotted lines to represent different colors.

1. Construct an equilateral triangle and color each side a different color. The figure will be easier to see if you choose **Display: Line Width: Thick**. Construct the interior.

2. Construct the midpoint of each side and the line segments connecting each vertex and the midpoint of the opposite side. Make these thin lines. These are the possible lines of reflection.

3. Select your original triangle and choose **Transform: Translate**. Remember that you can select multiple objects by holding down the mouse button while you move the mouse. In the **Translate** dialog box, select the option **Translation Vector: Rectangular**. After choosing this option, you will see two further choices: Fill in the **Horizontal** option by selecting **Fixed Distance**, and fill in the distance you wish to translate the triangle (you will see on the screen a ghost image, so use this to help decide how far to translate the image so it doesn't overlap the original triangle). Fill in the **Vertical** option with **Fixed Distance** and enter a distance of 0. Now click the **Translate** button at the bottom of the window. You will work from the translated triangle. The original triangle is the first in your series. Label it 1 as the identity.

4. In the translated triangle, construct the point of intersection for two of the lines inside the triangle. Mark it as the center of rotation by double-clicking on it or by selecting it and choosing **Transform: Mark Center**.

5. Select the second triangle and rotate it 120° by choosing **Transform: Rotate** and typing **120** in the box. Note that the coloring on this configuration is different from the previous one. This triangle is the second in your series. Keeping the second triangle selected, choose **Transform: Translate**. This gives you a third copy on which to continue working. Label the second triangle R for rotation.

6. Mark the center of the translated triangle as the center of rotation, and then rotate the figure by another 120° degrees. Note that this configuration is different from the previous two. Translate this triangle for a working copy, and label the third in your series R^2 for two rotations.

7. Mark the center and rotate the translated triangle by 120° degrees a third time. Note that this configuration is identical in shape and color to the original triangle and so does not need to be added into the picture series. Delete it.

1

R

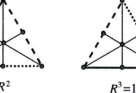

R^2 $R^3=1$

8. Now that we have pictures of all the rotations, we need to work on reflection. We will use the vertical line for our axis of reflection as in <u>Symmetry, Shape, and Space</u>, but first we will set up the triangles. Select the original triangle. Use the **Copy** and **Paste** commands in the **Edit** menu to create a copy. Then move the copy directly below the original. Keeping the new copy selected, translate it twice so you have a row of three copies of the original triangle on which to perform the rotate/reflect operations.

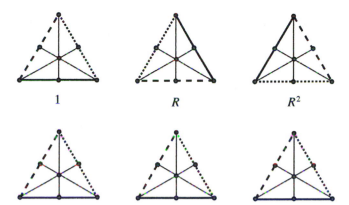

9. Double-click on the vertical line through the first triangle in the second row, or select it and choose **Transform: Mark Mirror**.

10. Select the triangle and its vertices and choose **Transform: Reflect**. Note that the coloring on this configuration is different from the previous ones. This triangle is the fourth in your series. Label it F for flip.

11. Now make the vertical line in the second triangle the mirror line and the center the rotation point. Rotate the triangle by $120°$ and reflect it. This configuration is RF. Order is important: Be sure to do the rotation first.

12. Move to the third triangle. Mark its vertical line and center as the mirror line and rotation point. Rotate the triangle by $120°$ twice and then reflect it. This configuration is $R^2 F$.

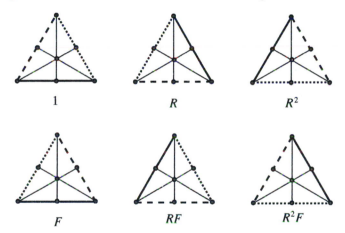

Now you have a series of pictures representing all distinct configurations for the triangle. Keeping that sketch open in the background, open a new sketch and copy and paste the original configuration (labelled 1) onto the new page for the following exercises. Note a few oddities in the software. Since $R^3 = 1$, *The Geometer's Sketchpad* can't imagine that you would want to do three rotations. It will not let you do a rotation back to the original position unless you translate your picture and work on the translated copy. The same is true with reflections. For example, mark the center of your new triangle as the mirror line. Select the triangle and choose **Transform: Reflect**. Note that the sides switch places as they should. If you try to reflect the triangle a second time, however, the sides will not change. Instead, choose **Transform: Translate** and then **Transform: Reflect**. Now the sides will be back in the original orientation, demonstrating that $F^2 = 1$. As long as you do not have three consecutive rotations for the triangle or two consecutive flips, you can use your triangle to simplify complicated expressions. Simply perform the operations in order on a new copy of the original triangle, and compare the resulting figure to your table of six configurations.

Of course, all of the computations of the exercises below are easier using the multiplication table as in Symmetry, Shape, and Space, but visual thinkers may prefer this more direct approach. The manipulations also lend credence to the computations from the tables.

▷ **Exercise 8.** Use your triangle to verify the following:
(a) $FR = R^2 F$
(b) $FR^2 = RF$

▷ **Exercise 9.** Use your triangle to simplify the following expressions to one of the six in the original group.
(a) $FR^2 FR$
(b) $RFRFRFRF$
(c) $FR^2 FRFR^2 FRF$
(d) $RFR^2 FR^2 FRFR^2 F$

▷ **Exercise 10.** Verify that reflecting about the other two mirror lines does not give a new configuration by marking each diagonal line as the mirror line, reflecting the triangle, and finding the figure from the original set of six pictures that it matches.

▷ **Exercise 11.** Draw a square, color the sides, and follow the outline of the previous demonstration to construct and label a series of eight pictures representing the distinct configurations of the square.

▷ **Exercise 12.** Make a copy of your original square to simplify the following:
(a) $FRFR^2 FR^3 FR$
(b) $FR^3 FR^2 F$
(c) $RFR^3 FRFR^2 F$

▷ **Exercise 13.** Verify that reflecting about the other three mirror lines of the square does not give a new configuration by marking each as the mirror line, reflecting the square, and finding the figure from the original set of eight pictures of Exercise 11 that it matches.

17. Frieze Patterns

Companion to Chapter 5.3 of <u>Symmetry, Shape, and Space</u>

To generate frieze patterns using *The Geometer's Sketchpad*, recall that all such patterns are generated using reflections (either across the line running down the center of the pattern or across any line perpendicular to this center line), rotations of 180° only, glide reflections along the center line, and translations in the direction of the center line.

Demonstration: Motif for a Frieze Pattern [★ Friezes. gsp]

1. Begin by drawing the center line *AB* of the hypothetical strip for the border pattern. Use **Construct: Line**, and set the two points *A* and *B* on the line at a distance that will form your basic translation.

2. Draw a basic asymmetric motif, which may lie on one side of the center line or may intersect it.

Demonstration: Frieze Pattern p111 or Hop

1. To draw a pattern of type *hop* or *p*111, we use only translations. There are two ways of entering the desired translation: using either coordinates or vectors. In this context, we think vectors are easier. Select points *A* and *B* in order, and choose **Transform: Mark Vector**.

2. Select your motif and choose **Transform: Translate**. A new screen will pop up, labelled **Translate**. Make sure that the option **Translation Vector: Marked** is chosen and that the boxes below say **From Point A** and **To Point B**. Click the **Translate** button, and a new copy of the figure will appear at the correct distance along the line.

3. Repeat to get as many copies of the figure as you like.

4. You can, at the end, hide the center line and the points *A* and *B*.

▷ **Exercise 1.** Follow the suggestions above to draw a *hop* or *p*111 pattern. Experiment with dragging the original motif and also with moving the points *A* and *B*.

Demonstration: Frieze Pattern p1m1 or Jump

1. To create a *jump* or *p*1*m*1 pattern, one can either begin with a hop pattern and reflect the whole pattern across the center line, or begin with a single motif, reflect it across the center line, and then translate the doubled motif along the vector from *A* to *B*.

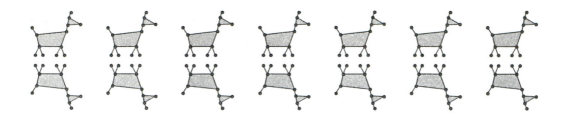

▷ **Exercise 2.** Follow the suggestions above to draw a *jump* or *p1m1* pattern two different ways. Experiment with dragging the original motif and also with moving the points *A* and *B*.

▷ **Exercise 3.** Beginning with a single asymmetric motif, figure out how to draw a *sidle* or *pm11* pattern. (You may need to change the spacing of *A* and *B*.)

▷ **Exercise 4.** Beginning with a single asymmetric motif, figure out how to draw a *spinning jump* or *pmm2* pattern.

Demonstration: Frieze Pattern p112 or Spinning Hop

1. To draw a *spinning hop* or *p112* pattern, we begin with our basic motif. Select point *B* and choose **Transform: Mark Center**.

2. Select your motif and point *A*, and choose **Transform: Rotate**. Fill in the angle as 180° in the **Rotate** window, and a new rotated figure will appear along the center line.

3. Select the new farthest-right point for your rotation center and repeat, being careful to include the last point created along the center line so that you'll have a new rotation center to work with.

4. Continue to form a line of figures. Note that, after you are done, you can select and drag the original motif or the points *A* and *B* to alter the whole line of figures to pleasing effect. Go back and hide the center line and the points along it.

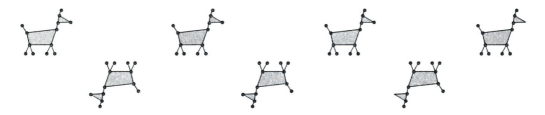

▷ **Exercise 5.** Follow the directions above to draw a *spinning hop* or *p112* pattern. Experiment with dragging the original motif and also with moving the points *A* and *B*.

▷ **Exercise 6.** Beginning with a single asymmetric motif, figure out how to draw a *spinning sidle* or *pmg2* pattern.

Demonstration: Frieze Pattern p1g1 or Step

Unfortunately, *The Geometer's Sketchpad* does not include the glide reflection operation as one of the basic commands. However, it is easy to generate a *step* or *p1g1* pattern using reflections and translations.

1. One method is to take a *jump* or *p1m1* pattern already created and hide every other image, using **Display: Hide Objects**.

2. Another way, starting from a single motif, is to first reflect the image over the center line, and then translate it along the vector from *A* to *B*. Hide the reflected motif, and then translate the remaining pair along the center line.

▷ **Exercise 7.** Follow the suggestions above to draw a *step* or *p1g1* pattern. Experiment with dragging the original motif and also with moving the points A and B.

For more advanced users, *The Geometer's Sketchpad* includes a **Custom Tool** option which we can use to generate glide reflections. To use this, perform the operation, then select in order the starting configuration (the inputs) and then the desired outcome (the outputs). Note that the inputs should consist of the defining objects. For example, only input points A and B rather than the points and the line AB which is defined by the points. [★ Glide Reflection Tool.gsp]

18. Wallpaper Patterns

Companion to Chapter 5.4 of <u>Symmetry, Shape, and Space</u>

The seventeen wallpaper patterns can be classified by their combinations of lines of reflection, lines of glide reflection, and rotation centers of various orders, all of which can be laid out nicely with *The Geometer's Sketchpad*. First draw an asymmetric motif to use for all of your wallpaper patterns. Color lines of reflection red and lines of glide reflection blue to make them easier to identify. We begin with the pattern **p111**, which has only translations on a grid of parallelograms. [★ Wallpaper.gsp]

Demonstration: Wallpaper Pattern p111

1. Draw a horizontal line segment with two points A and B indicating the distance of the first translation.

2. Draw another line segment AC at an angle to AB. Insert your motif within the angle.

3. Select first A and then B, and click on **Transform: Mark Vector**.

4. Select your motif and choose **Transform: Translate**. The pop-up window labeled **Translate** should say **Translation Vector: Marked**.

5. Repeat to get enough copies of the motif to fill a row stretching across your window.

6. Now select points A and C in order and choose **Transform: Mark Vector**.

7. Select the row of motifs and translate in that direction. Repeat to fill up the screen, and you will end up with part of a wallpaper pattern of type **p111**.

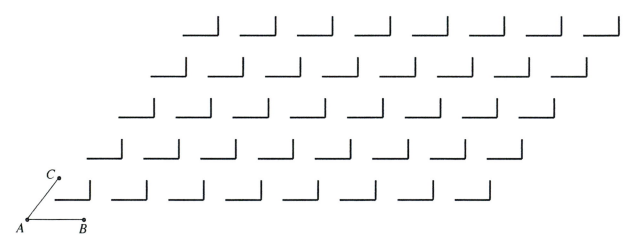

▷ **Exercise 1.** Draw a **p111** pattern with your own motif.

▷ **Exercise 2.** Create a new copy of your *Sketchpad* sketch for Exercise 1, either by copying and pasting to a new sketch or by using **File: Document Options: Add Page: Duplicate: 1**. Delete all but the original motif and the line segments AB and AC. Place a point near the center of the parallelogram outlined by AB and AC. Use this as a center of rotation and rotate the motif by 180°. Now translate the doubled motif, first in the direction of AB and then in the direction of AC, to get the pattern **p112**.

Next, we'll draw the wallpaper patterns that use a rectangular grid: **p1m1**, **p1g1**, **p2mm**, **p2mg**, and **p2gg**. To draw all of these, begin constructing a right angle as in Steps 1–3 of the demonstration below for **p1m1**. Place your motif within this angle.

Demonstration: Wallpaper Pattern p1m1

1. Draw a horizontal line segment AB.

2. Select the point A and the segment, and use the **Construct: Perpendicular Line** command to construct a line perpendicular to AB.

3. Place point C on this line so that AC is a convenient distance to translate your motif in the vertical direction. Place your motif within the angle formed by C, A, and B.

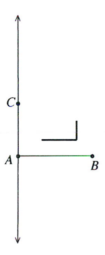

4. To make the wallpaper pattern **p1m1**, we need reflections in one direction. Select the line AC.

5. Choose the **Transform: Mark Mirror** option. Note that double-clicking on the line so that it flashes will also mark it as a mirror.

6. Now select your motif and choose **Transform: Reflect** to get its mirror image.

7. Select first point A and then C, and choose **Transform: Mark Vector**.

8. Translate the doubled motif in the direction of AC to fill up your screen.

9. Once you have a stack of double motifs, select first A and then B, and choose **Transform: Mark Vector**.

10. Translate the whole stack of double motifs in the AB direction several times. If the motifs overlap, don't worry since you can fix that later. Once you have the patterns laid out, adjust the spacing of A, B, and C to make a pleasing wallpaper pattern. Try fiddling with each of these points in turn.

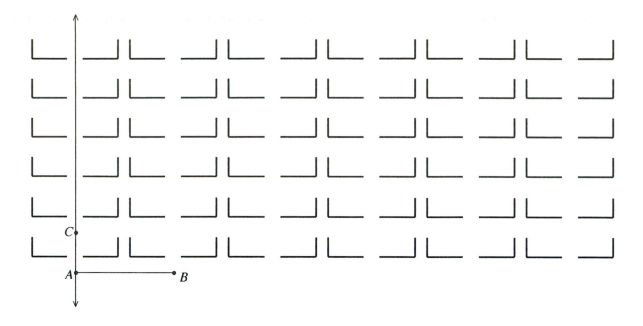

▷ **Exercise 3.** Draw a **p1m1** pattern with your own motif.

Demonstration: Wallpaper Pattern p1g1

1. We can modify the **p1m1** pattern to make a **p1g1** pattern. To keep the original wallpaper pattern intact, either copy the **p1m1** pattern to a new sketch or add a new page duplicating the pattern using the **File: Document Options: Add Page** option.

2. Delete all but the first stack of doubled motifs.

3. Choose every other one and use **Display: Hide Objects** to hide them. Do not delete them or their reflections and translations will also be deleted.

4. Now choose translation vector AB and translate your new stack to fill up the screen. Tinker with the spacing of A, B, and C until it looks nice.

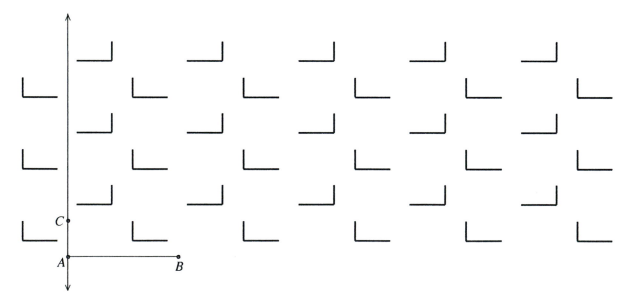

▷ **Exercise 4.** Draw a **p1g1** pattern with your own motif.

▷ **Exercise 5.** Below is a section of **p2mm** wallpaper. Figure out how to draw this using *The Geometer's Sketchpad*, and make such a pattern using your own motif.

▷ **Exercise 6.** Below is a section of **p2mg** wallpaper. Figure out how to draw this using *The Geometer's Sketchpad*, and make such a pattern using your own motif.

▷ **Exercise 7.** Below is a section of **p2gg** wallpaper. Figure out how to draw this using *The Geometer's Sketchpad*, and make such a pattern using your own motif.

▷ **Exercise 8.** Below is a section of **c2mm** wallpaper. Figure out how to draw this using *The Geometer's Sketchpad*, and make such a pattern using your own motif.

▷ **Exercise 9.** Below is a section of **c1m1** wallpaper. Figure out how to draw this using *The Geometer's Sketchpad*, and make such a pattern using your own motif. [Hint: It is easiest to start by modifying your **c2mm** pattern.]

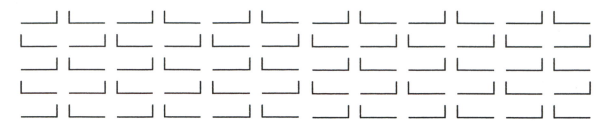

The next group of patterns, **p411**, **p4mm**, and **p4gm**, have four-fold rotations of 90° and use a square lattice. Set up the lattice by following Steps 1–3 of the demonstration below for **p411**.

Demonstration: Wallpaper Pattern p411

1. Draw a line segment AB and construct a line through A perpendicular to AB.
2. Draw a circle centered at A with radius AB, and place point C at the intersection of the circle and the perpendicular line. Hide the circle using **Display: Hide Circle**.
3. Place your motif inside the angle formed by C, A, and B.
4. To generate the pattern **p411**, choose point A as a rotation center (using **Transform: Mark Center** or by double-clicking on A so it flashes), and rotate your motif by 90°.
5. Repeat to get four copies of the motif around point A.
6. Select first point A and then B, and choose **Transform: Mark Vector**.
7. Select all four copies of the motif, and translate this group in the AB direction to fill one row across your screen. Adjust the spacing of A and B if necessary.
8. Select first A and then C, mark this as a translation vector using **Transform: Mark Vector**, and translate the row of quadruple motifs in the AC direction.

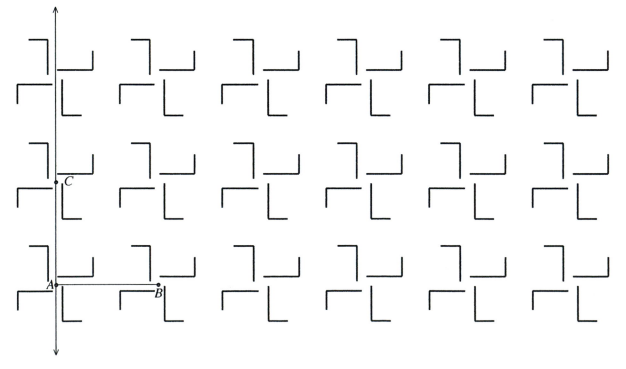

▷ **Exercise 10.** Draw a **p411** pattern with your own motif.

▷ **Exercise 11.** To draw a **p4mm** pattern, make a copy of your **p411** pattern, and delete everything except the lines *AB* and *AC* and the four copies of the rotated motif centered around *A*. Adjust the spacing of *A* and *B* so that your motifs are spread out. Select the segment *AB* or the line *AC* (either will do), and choose **Transform: Mark Mirror**. Choose all four copies of the motif, and reflect to get eight copies (with four lines of reflection). Now translate this group of eight in both directions to fill up your screen.

Demonstration: Wallpaper Pattern p4gm

1. The pattern **p4gm** is of course more trouble, since we have to do glide reflections by hand. It is probably easiest if one starts with a **p4mm** pattern and hides bits of it to make a **p4gm**. Below are pictures of each. The dashed lines indicate lines of reflection and the dotted lines are lines of glide reflection. Squares denote four-fold centers of rotation; they are solid if they occur along a line of reflection and open if not. Diamonds indicate two-fold centers of rotation. Modifying the **p4mm** pattern to turn it into **p4gm** does not remove any of the indicator lines or rotation centers, but it does change their type in many cases.

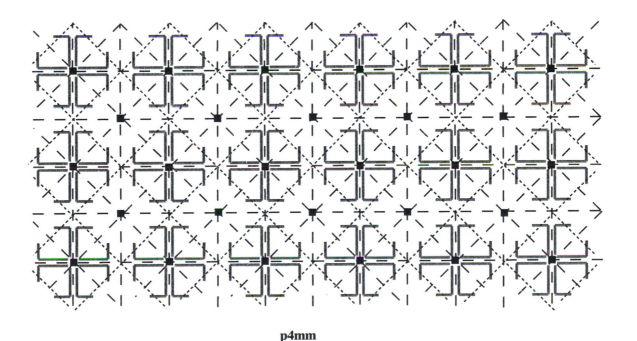

p4mm

91

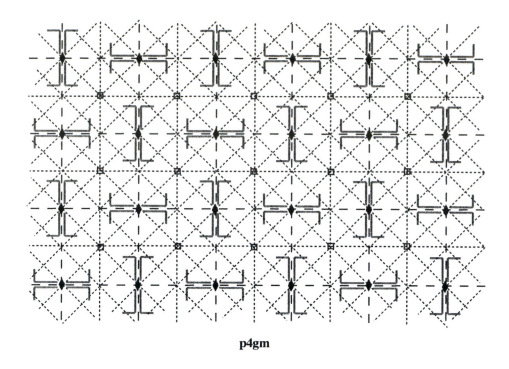

p4gm

▷ **Exercise 12.** Draw a **p4gm** pattern with your own motif.

We are left with the five patterns involving three-fold and six-fold rotations. The patterns **p311** and **p611** are easy: Do these just as we did **p411** above but use rotations of 120° and 60° instead of 90°. In this way you will translate along lines that form a 60° angle.

▷ **Exercise 13.** Draw a **p311** pattern with your own motif.

▷ **Exercise 14.** Draw a **p611** pattern with your own motif.

The most confusing of all the wallpaper patterns are **p31m** and **p3m1**. The difference between these two is the placement of the rotation centers: In **p3m1** all rotation centers are on the lines of reflection, while in **p31m** some are on the reflection lines and some are not. In the drawings below, the reflection lines are shown as dashed, and the rotation centers are represented by triangles, solid if on the lines of reflection and open if not. The lines of glide reflection have been omitted.

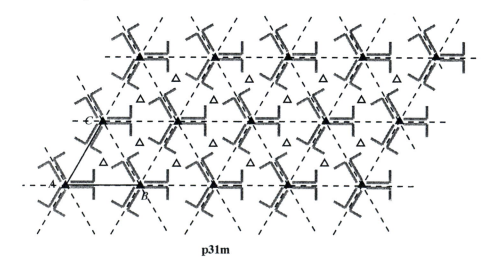

p31m

92

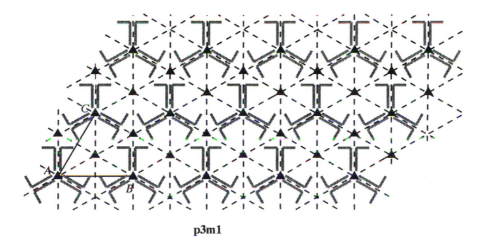

p3m1

A close examination of these pictures allows us to figure out how to draw them with *The Geometer's Sketchpad*.

Demonstration: Wallpaper Patterns p31m and p3m1.

1. To draw **p31m**, start with a horizontal line AB, and rotate it 60° about A to form another line AC. Place your motif in the angle.

2. Use **Transform: Mark Mirror** and **Transform: Reflect** to reflect the motif across AB.

3. Rotate the doubled motif by 120° about center A twice to get six copies of the motif (and three lines of reflection).

4. Translate this six-fold motif first in the AB direction to form one row of the pattern.

5. Translate this row in the AC direction to get a section of a **p31m** pattern.

6. To draw the quite similar **p3m1** pattern, repeat the process to get a horizontal segment AB and another segment AC forming a 60° angle.

7. As above, reflect and rotate to get the six-fold reflected group of motifs, but this time, before translating it, rotate the six-fold group 90° (or 30°) about center A. This ensures that the lines of symmetry of the basic group are not the same as the translation vectors.

8. Translate the six-fold group in the direction of AB to form one row of the pattern.

9. Translate this row in the AC direction to get a section of a **p3m1** pattern.

▷ **Exercise 15.** Draw a **p31m** pattern with your own motif.

▷ **Exercise 16.** Draw a **p3m1** pattern with your own motif.

The wallpaper pattern **p6mm** is complicated, in that it involves a lot of copies of the motif, but not that difficult to draw with *The Geometer's Sketchpad*. Model the procedure used for **p4mm**, but rotate the original motif and its reflection by 60°. Repeat to get 12 copies (six copies of the original plus six reflections) centered around A. Translate in the horizontal direction to get one row of motifs, and then translate the row in the direction that is 60° from the horizontal.

▷ **Exercise 17.** Draw a **p6mm** pattern with your own motif.

19. Islamic Lattice Patterns

Companion to Chapter 5.5 of <u>Symmetry, Shape, and Space</u>

As we mentioned in discussing celtic knots, *The Geometer's Sketchpad*, like most drawing programs, cannot handle the over-under weaving necessary for the best lattice patterns. However, it is an invaluable tool for laying out the basic design. It was used for this purpose for the drawings in <u>Symmetry, Shape, and Space</u> and then these preliminary drawings were exported into another program, *Macromedia*® *FreeHand*®, where the weaving pattern was introduced one intersection at a time, an extremely laborious process. For example, the large illustration on page 178 of the text took four hours after throwing away two previous attempts.

What *The Geometer's Sketchpad* does do beautifully is lay the rings of dots and then lay out the rings in a regular grid.

Demonstration: Islamic Lattice Patterns

Each lattice pattern is based on a grid of circles of dots. These can be generated as in Section 10: Star Polygons. The most popular choices are six, eight, or twelve dots, though you are quite welcome to try more if you like.

1. Place two dots, select one of them, and choose **Transform: Mark Center**.

2. Select the other dot, and choose **Transform: Rotate** to rotate it about the center by 60°, 45°, or 30°, depending on how many dots you want. In general, an angle of $\frac{360°}{n}$ will give n dots.

3. Repeat the rotation to get a circle of dots.

4. Use **Display: Hide Point** to hide the center point. Once you have a circle of dots, figure out how you want to space the circles. All lattice patterns can be classified as wallpaper patterns, and the designs are almost always of type **pxmm**, where **x** must be 2, 4, or 6. Type **p2mm** is rarely used, since higher orders of symmetry are generally preferred. For eight dots (or any other multiple of four), you can use pattern **p4mm**. For six or twelve dots, you can use **p6mm**.

5. Draw a horizontal line segment AB, which will be used as one of your translation vectors.

6. Rotate AB about A at an angle of 90° for **p4mm** or by 60° for **p6mm** for your other translation vector AC.

7. Translate your circle of dots in the AB direction to get a row of circles of dots.

8. Then translate this row in the AC direction several times.

9. Adjust the size of the rings by dragging one dot of the original ring. Adjust the spacing of the circles by dragging point B.

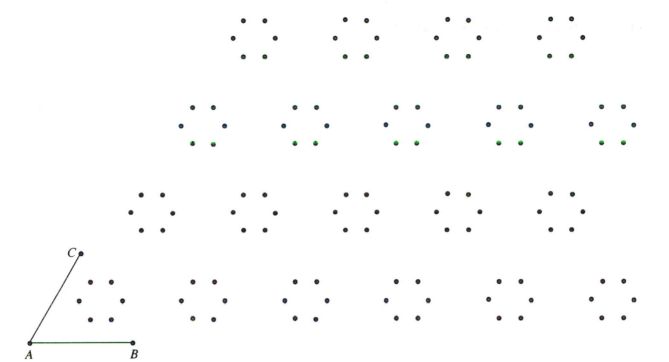

10. Now draw a line segment connecting one dot on one circle to another dot on another circle. Different choices here will give different patterns, some of which do not work out. Ideally (but not invariably in practice), the angles formed around each circle of dots will all be equal, so that a star pattern is formed around each circle. If we adopt this rule, then the first line drawn dictates all of the others. All of the other lines could be formed by reflecting this line across the diameter of the circle or by rotating the line segment to another dot on the circle. However, doing this with *The Geometer's Sketchpad* causes the lines either not to meet or to cross when we later change the size or spacing of the circles, since the line segments will not be tied at both ends to the dots on the circles. Therefore, you are faced with the task of figuring out what dot to go to next to keep the angles constant. However, once you have traced a chain of line segments of the design from dot to dot across the width or height of the grid of circles you have laid out, you can translate the chain instead of having to draw it again.

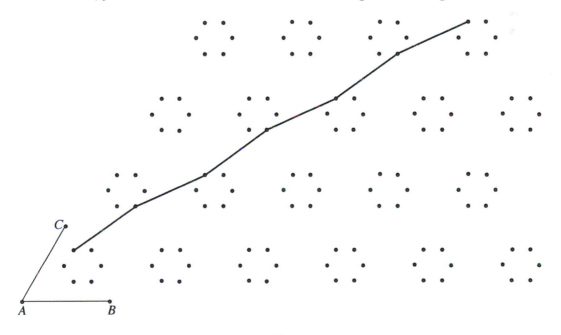

95

▷ **Exercise 1.** Lay out a grid of circles with six dots in a **p6mm** pattern. Connect the dots as shown below by drawing one copy of each chain of segments and translating it. Experiment with changing the spacing of the dots around the circles (by dragging one point of the original ring of dots) and with changing the spacing of the circles (by dragging the end B of the translation vector AB). [★ Lattices.gsp: Lattice #1]

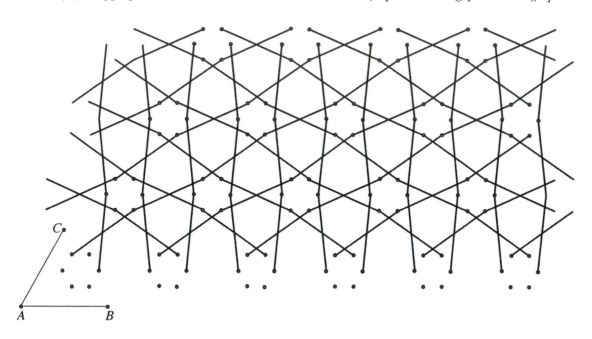

▷ **Exercise 2.** [**SSS 5.5.4 and 6**] Lay out a grid of circles with six dots in a **p6mm** pattern. Connect the dots as shown below by drawing one copy of each hexagonal chain of segments and translating it. Experiment with changing the spacing of the dots around the circles by dragging one dot of the original circle, and with changing the spacing of the circles by dragging the end of the translation vector AB. [★ Lattices.gsp: Lattice #2]

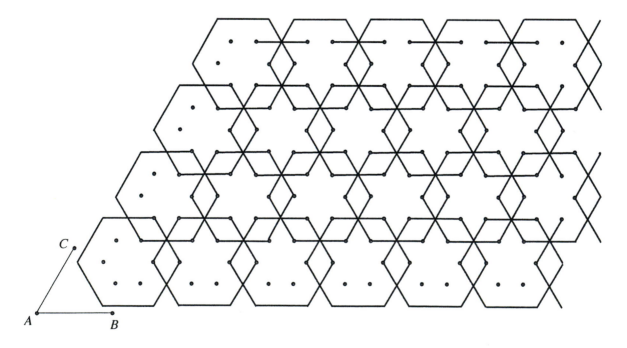

▷ **Exercise 3.** **[SSS 5.5.9]** Figure out the relationship between the size of the circles and their spacing in the figure for Exercise 2 on the previous page.

▷ **Exercise 4.** **[SSS 5.5.10]** Lay out a grid of circles with twelve dots in a **p6mm** pattern. Connect the dots as shown below by drawing one copy of each chain of segments. Finish the pattern by translating the chains. Experiment with changing the spacing of the dots around the circles and with changing the spacing of the circles. [★ Lattices.gsp: Lattice #3]

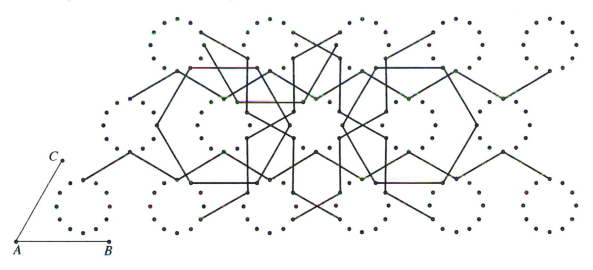

▷ **Exercise 5.** **[SSS 5.5.13]** Here is a more complicated arrangement with two sizes of circles and three arrangements of dots: larger twelve-dot circles and small two- and three-dot circles. Connect the dots as shown below by drawing one copy of each chain of segments and translating it. Experiment with changing the spacing of the dots around the circles and with changing the spacing of the circles. [★ Lattices.gsp: Lattice #4]

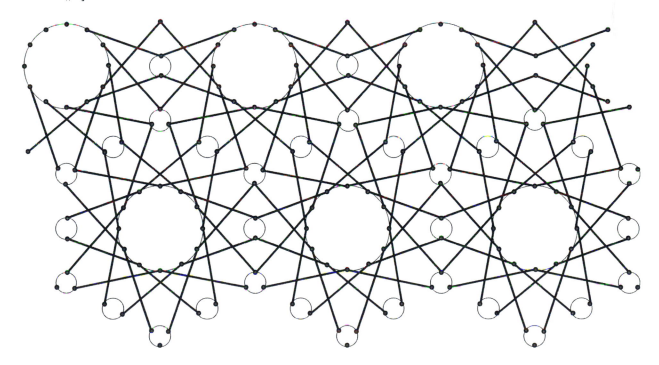

20. The Hypercube

Companion to Chapter 6.2 of <u>Symmetry, Shape, and Space</u>

In this brief section we explain how to model a four-dimensional hypercube with *The Geometer's Sketchpad*. Recall that a point is considered to be zero-dimensional. Dragging a point one unit to the east (or west) gives a one-dimensional unit-long line segment. Moving east and west defines one dimension. Dragging a line segment one unit north (or south) gives a two-dimensional square. Moving north and south in a direction perpendicular to the east-west direction defines a second dimension. Now if you take the square, lay it down on the ground, and move it up and down (in a direction perpendicular to the plane defined by east-west and north-south), this gives a three-dimensional cube. In the figure below on the right, one could also describe the x direction as forward-back, y as left-right, and z as up-down.

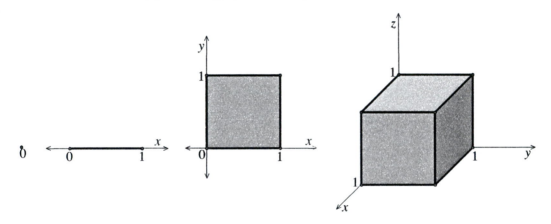

The Geometer's Sketchpad is rooted in the second dimension, as are the pieces of paper and the blackboard you have used in class. The third dimension, perfectly intelligible to anyone who can see you waving your hands around, requires some fakery to represent on paper (or on the computer screen). The standard mathematical convention is to represent the third direction as a skew line.

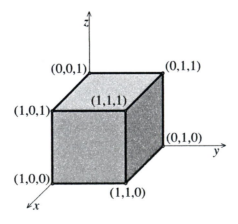

It is to be understood that the line representing the x-axis is perpendicular to the y- and z-axes, even if it doesn't look like it in the picture. This idea leads us to the realization that we can represent the fourth dimension, what C.H. Hinton called the *ana-kata* direction, as yet another skew line on the piece of paper (or computer screen). We must keep in mind that this fourth direction actually points in some unseeable direction perpendicular to all three of the other directions. As long as you don't let that bother you, we can proceed to draw the hypercube.

Demonstration: Hypercube [★ Hypercube.gsp]

1. Draw a horizontal line segment, and label this line segment y.

2. Construct a line through the left endpoint of this line segment and perpendicular to it.

3. Place a point on this line, draw the line segment from this point to the point of intersection of the perpendicular line and the initial line segment, and then hide the line. Label this perpendicular line segment z.

4. Draw a third line segment from the point of intersection, and label it x.

5. Draw a fourth line segment from the point of intersection, and label it w. The y and z lines represent the two dimensions: left-right and up-down. These are constrained to be perpendicular. The x-direction represents forward and backward and is understood to be perpendicular to y and z. The w-direction represents the *ana* and *kata* directions and is understood to be perpendicular to the three others.

6. Color your line segments four different colors.

7. Place a point to one side of the axis system just created. Click on first the left endpoint and then the right endpoint of line segment y, and choose **Transform: Mark Vector**.

8. Select your new point of Step 7, and move it in the y direction using **Transform: Translate**.

9. Connect these two new points with a line segment (using either the **Segment** tool or **Construct: Segment**), and color the segment with the y-color.

10. Select first the lower endpoint of z and then the upper endpoint, and choose **Transform: Mark Vector**.

11. Select the line segment constructed in Step 9 and its endpoints, and translate it in the z direction. You now have parallel line segments.

12. Connect corresponding endpoints with segments, and color these segments with the z-color. You now have a hollow square.

13. Select the endpoints of the x line segment and choose **Transform: Mark Vector**.

14. Select the square and translate it in the x direction.

15. Connect corresponding endpoints of the two squares with segments, and color these segments with the x-color. You now have a wire-frame cube.

16. Select the endpoints of the w line segment, and choose **Transform: Mark Vector**.

17. Select the cube and translate it in the w direction.

18. Connect corresponding endpoints of the two cubes with segments, and color these segments with the w-color. You now have a wire-frame hypercube.

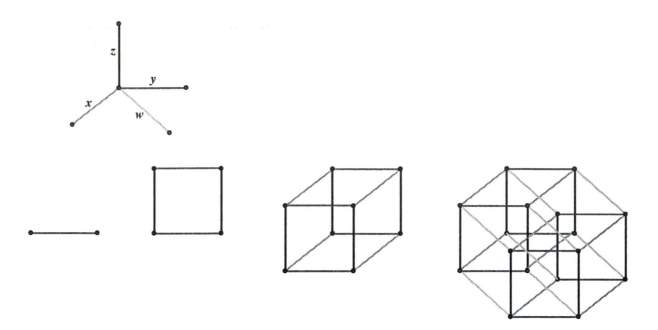

All of the steps of this process are illustrated above, though we are only interested in the original axis system and the final hypercube. What is more, this hypercube can be manipulated by dragging the ends of the axes to see the effect of changing the scale on each axis and changing the angles representing z and w. Drag the free endpoints of each axis, and observe the effect on the hypercube drawing.

▷ **Exercise 1.** Follow the instructions above to make your own hypercube, and play with it.

You can name locations or points in any dimension. On the line segment, for example, we can represent a point as a number that measures distance in the the usual positive sense from the endpoint we choose as 0. Our thanks to Helen Gerretson for suggesting the following series of exercises.

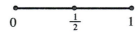

$$0 \qquad \frac{1}{2} \qquad 1$$

Likewise, we can label points in the square using two coordinates (x, y). The first number x indicates the east-west distance, and the second number y indicates the north-south distance.

▷ **Exercise 2.** Find the coordinates of the points C, D, E, and F in the picture below. Note that E is positioned at the center of the square.

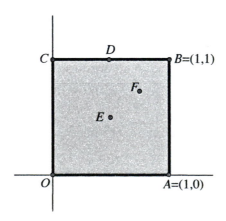

100

▷ **Exercise 3.** (a) Label all the vertices of the unit cube (the cube that has edges of length 1).
(b) Note that the vertices of the cube all consist of triples with 0's and 1's only. What figure is formed if you connect only the vertices with an even number of 1's?

▷ **Exercise 4.** (a) Label all the vertices of the hypercube of side length 1.
(b) Plot the point $(\frac{1}{2}, \frac{1}{2}, \frac{1}{2}, \frac{1}{2})$.

21. Polyhedra

Companion to Chapters 7.2 and 7.3 of <u>Symmetry, Shape, and Space</u>

The Geometer's Sketchpad produces sketches in two dimensions, so the easiest use of it in the study of polyhedra is in the drawing of two-dimensional representations of the solids. There are three main forms of these representations: perspective drawings (see Section 24), nets, and Schlegel diagrams. We will concentrate on drawing nets of the polyhedra. The Schläfli symbol, such as 4.4.4 for the cube, gives some indication of the types of faces and the vertex configuration.

Recall that a net for a polyhedron is a two-dimensional figure which could be cut out and reassembled to form the polyhedron. Nets must join the polygonal faces along their edges and must be connected. Nice models of polyhedra can be constructed by printing the net on stiff paper or cardboard, scoring the edges of each polygon with a pen tip or knife, and taping the edges together. Below is one net for the cube:

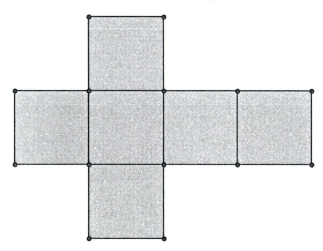

Demonstration: Net for the Cube 4.4.4 [★ Platonic Solids.gsp]

The net above can be constructed in the following manner. First, construct a square. This can be done by standard ruler and compass techniques: Draw a line segment, erect perpendicular lines at both endpoints, draw circles centered at each endpoint with radius equal to the segment, mark the intersections with the perpendicular lines, connect the vertices with segments, construct and color the interior, and then hide the circles and the long lines. Alternatively, use either Method 2 or 3 of Section 3. Of course, if a polygon of seven or nine sides is required, we must use Method 2 or 3, since these are not constructible by ruler and compass. Rather than constructing each of the six squares of the net in this manner, we only construct one and then generate copies using the **Transform** menu. For example, assume that you have constructed a single square, and that now you want to get a new square adjoining the old one along the right-hand edge. There are three ways of doing this.

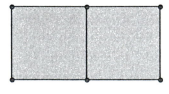

102

1. The first way involves translation. Select in turn the upper left vertex and then the upper right vertex, and choose **Transform: Mark Vector**. Then select the entire square (you can omit the left-hand edge and its vertices if you're fussy), and choose **Transform: Translate**. In the box that pops up, make sure that **Translation Vector: Marked** is chosen, and hit the **Translate** button.

2. A second way of generating the second square is by reflection. Select the right-hand edge, and choose **Transform: Mark Mirror** (or double-click on the edge so that it flashes). Then select the square (omitting the right-hand edge and its vertices if you like), and choose **Transform: Reflect**.

3. The third way to get a second square in the desired position is by rotation. Select the upper right vertex, and choose **Transform: Mark Center** (or double-click on the vertex so that it flashes). Then select the square (omitting the top edge and its vertices if you wish), and choose **Transform: Rotate**. Make sure that the option **Translate by Fixed Angle** is checked. The default angle of rotation is 90°, which in this case is just what we want, so click the **Rotate** button. For a more complex polygon, you can compute the angle of rotation by recalling the formula $\frac{(n-2)180°}{n}$ for the vertex angle of an n-sided polygon. If you don't want to use formulae, you can mark the angle of the desired rotation by selecting three adjacent vertices of the polygon and choosing **Transform: Mark Angle**. In the **Rotate** pop-up box it should then say **by Marked Angle**. Click on the **Rotate** button.

All three methods work in this case, and they are approximately equal in complexity.

▷ **Exercise 1.** Draw a regular hexagon (by ruler and compass techniques, Method 2, or Method 3). Generate an adjoining hexagon by:
(a) translation
(b) reflection
(c) rotation

▷ **Exercise 2.** Draw two nets for the cube, both different from the example above and from each other.

▷ **Exercise 3.** Draw three different nets for the tetrahedron.

Some nets for simple figures can be figured out by simply envisioning how the polygons connect. Another way of finding a net for a polyhedron is to take a cardboard model and cut it open until it lies flat. Make sure that your net remains connected and that polygons are joined along edges, rather than only at a vertex.

▷ **Exercise 4.** Draw two different nets for the octahedron.

Polyhedra such as the prisms, the pyramids, and the Archimedean solids have more than one type of polygonal face. For these, a variation of the general technique is used. It always helps to have a clear picture of what you are trying to draw. Here are step-by-step instructions for how we drew a net for the

103

cuboctahedron, using a hodgepodge of techniques. The polygons are numbered to make it easier to explain the stages. Remember that there are a lot of different nets and many ways to generate the polygons. [★ Archimedean Solids.gsp]

Demonstration: Net for the Cuboctahedron 3.4.3.4

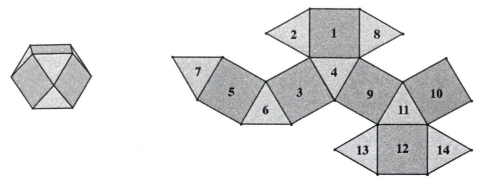

1. Draw Square **1**, using any method. Construct and color the interior.
2. Select the lower left vertex and choose **Transform: Mark Center**. Select the left edge and the upper left vertex, and choose **Transform: Rotate**. Type **60** for the angle.
3. Select the new endpoint of the line segment just generated, and choose **Transform: Mark Center**. Select this line segment and its other vertex, and choose **Transform: Rotate** to rotate by 60°.
4. Choose the three vertices of Triangle **2** just formed, and choose **Construct: Triangle Interior**. Change the color using **Display: Color**.

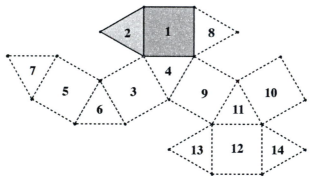

5. Now select the lower left corner of Square **1**, and choose **Transform: Mark Center**. We want to rotate the square into the position of Square **3**, so the square must rotate around itself (90°) and then around the triangle at position **4** (60°). All of these angles are measured in the clockwise direction, which counts as negative. Select Square **1**, and rotate by −150° using **Transform: Rotate**.
6. Draw the line segment that forms Triangle **4** using the existing vertices of Squares **1** and **3**. Select the three vertices, and construct and color the interior of the triangle.

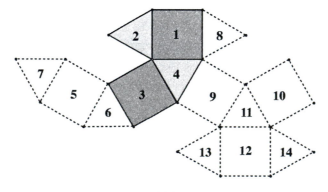

7. Rotate Square **3** about its upper left vertex through an angle of $-150°$ to position **5**.

8. Form Triangle **6** as in Step 6.

9. Form Triangle **7** as in Steps 2 through 4.

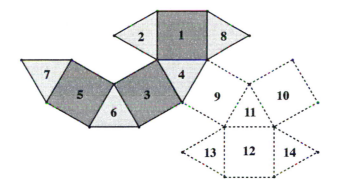

10. Select the lower right vertex of Square **1**, and choose **Transform: Mark Center**. Select Triangle 4, and rotate by $-150°$ to position **8**.

11. Select first the upper left vertex of Square **5** and then the lower right vertex of Square **1**. Choose **Transform: Mark Vector**. Select Square **5**, Triangle **6**, and Square **3**, and choose **Transform: Translate** to create new squares and a triangle at positions **9**, **10**, and **11**.

12. Select first the upper left vertex of Square **1** and then the lower left vertex of Triangle **11**, and choose **Transform: Mark Vector**. Select Square **1** and Triangles **2** and **8**, and translate them into positions **12**, **13**, and **14**.

▷ **Exercise 5.** Draw a net for the right regular square pyramid. First, construct a square and an adjoining triangle as in Steps 1–4 of the preceding demonstration. Then, rotate copies of the triangle into position on the other three sides of the square.

▷ **Exercise 6.** Draw a net for a right regular triangular prism. First, construct an equilateral triangle and an adjoining square on one of the edges of the triangle. Then, rotate copies of the square into position on the other two sides of the triangle. Add another triangle to make the top face of the prism.

▷ **Exercise 7.** Draw a net for the dodecahedron (twelve pentagonal faces with three meeting at each vertex). The Schläfli symbol for the dodecahedron is 5.5.5, indicating that three pentagons should meet at each vertex.

▷ **Exercise 8.** Draw a net for the rhombicuboctahedron 3.4.4.4.

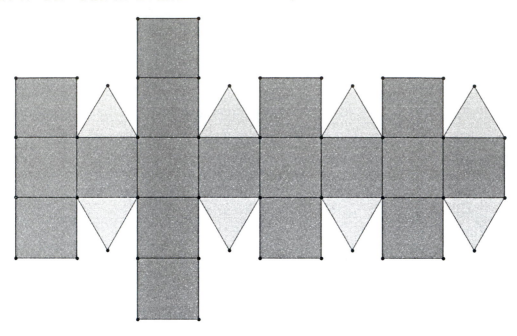

▷ **Exercise 9.** Draw a net for the snub cube 3.3.3.3.4.

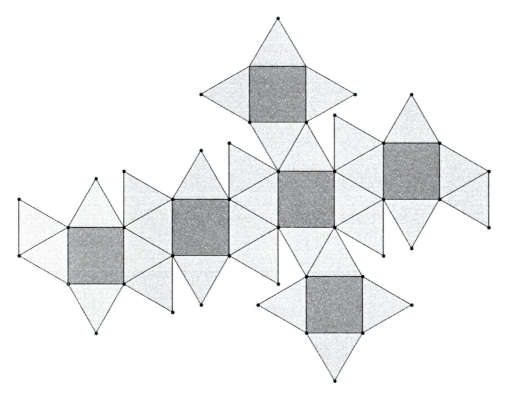

22. Spirals

Companion to Chapter 9.1 of <u>Symmetry, Shape, and Space</u>

The Geometer's Sketchpad can help you draw beautiful spirals. Depending on the method of construction, some are very precise and others are approximations. You may have already constructed versions of the Pythagorean spiral (Section 6) and the golden spiral (Section 7). The golden spiral in Section 7 was constructed using golden rectangles. It can also be constructed around the nested sequence of golden triangles constructed there.

A rectangular Archimedean spiral is the easiest to construct by hand or by mouse. [★ Spirals.gsp: Rectangular Archimedean]

▷ **Exercise 1.** Choose **Graph: Grid Form: Square Grid**. Draw a rectangular Archimedean spiral as shown below. Choosing **Graph: Snap Points** will guarantee that the line segments begin and end at integer points. Choosing **Display: Line Width: Thick** will make the spiral easier to see.

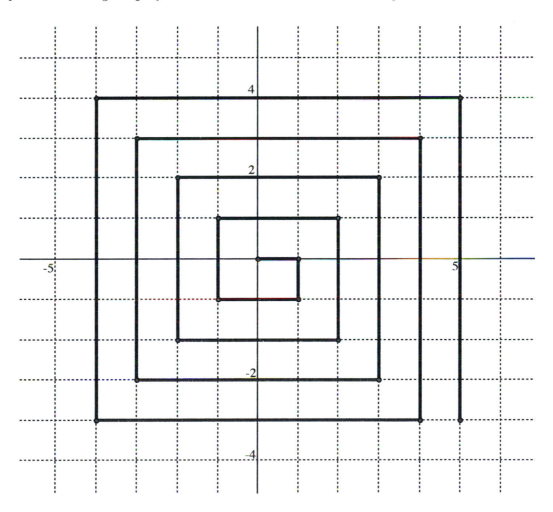

Next we construct the logarithmic spiral generated by the self-gnomon rectangle. Recall from <u>Symmetry, Shape, and Space</u> that two copies of the rectangle must make a larger rectangle with similar proportions. The sides must be in a $1 : \sqrt{2}$ ratio.

▷ **Exercise 2.** Construct a rectangle with length $\sqrt{2}$ and height 1.

Demonstration: Constructing a Logarithmic Spiral [★ Spirals.gsp: Approximate logarithmic]

1. Construct a rectangle with length $\sqrt{2}$ and height 1 as in Exercise 2.

2. Double-clicking on a side of a rectangle will mark it as the mirror line, or one can select the side of the rectangle and use **Transform: Mark Mirror**. Selecting the desired rectangle and choosing **Transform: Reflect** will create a second copy. Use this command repeatedly to generate the picture below and a few more iterations.

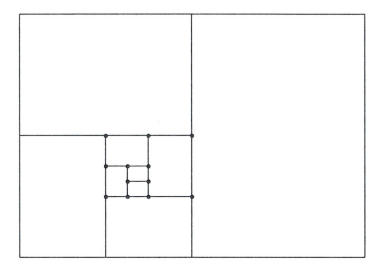

3. Use this set of rectangles to construct the polygonal spiral as shown below. [★ Spirals.gsp: Polygonal logarithmic]

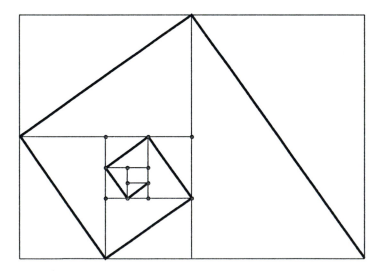

4. You can approximate the spiral curve by selecting three points in order, and then choosing **Construct: Arc Through 3 Points**. Then select the last of those points and the following two, and construct another arc, and so on. Note that this is only an approximation and that you would get a different approximation by choosing three different sequential points to define the arc.

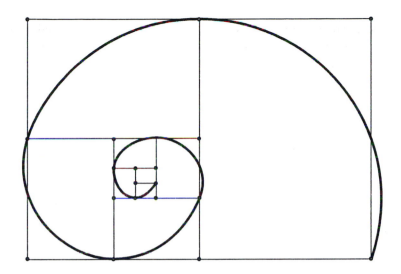

▷ **Exercise 3.** Use the rectangle constructed in Exercise 2 and the **Transform: Reflect** command to generate first the set of rectangles of increasing size, then the polygonal spiral, and finally the curved approximation of the logarithmic spiral.

▷ **Exercise 4.** [SSS 9.1.11] Use the **Graph: Grid Form: Square Grid** command, and then draw a polygonal logarithmic spiral as follows: At the center of your grid, draw a line segment 1 unit long. At the end of that segment and turned 90° clockwise, draw a segment 2 units long. At the end of that segment and turned 90° clockwise, draw a segment 4 units long. Turn 90° and draw a segment 8 units long. Continue the process, making each segment twice as long as the previous one, to the end of your sketch. Resize the grid smaller, and continue as far as you can. Use the process described above (of choosing three points in order and using **Construct: Arc Through 3 Points**) to connect the corners of your polygonal spiral with arcs.

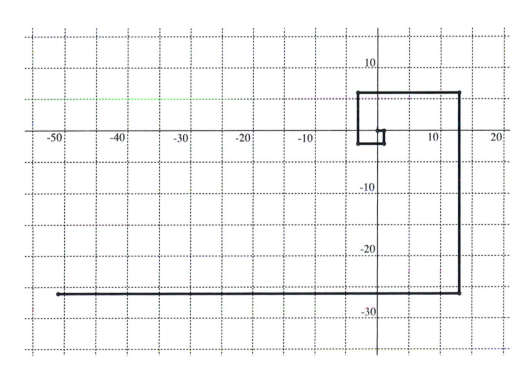

▷ **Exercise 5.** **[SSS 9.1.12]** Use the fact that the limit of the Fibonacci ratios is equal to the golden ratio to approximate the golden spiral with a Fibonacci spiral. Draw squares with side lengths equal to the Fibonacci numbers, and connect the corners with arcs of quarter circles (using the **Construct: Circle by Center + Point** and **Construct: Arc on Circle** commands).

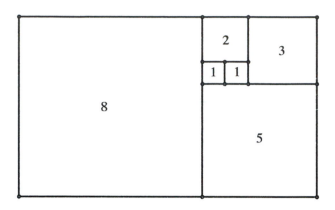

Graphing precise spiral curves is easiest in polar coordinates. Most spiral equations are given in polar form, and *The Geometer's Sketchpad* will let you define an equation and graph it in this coordinate system. We will begin by introducing the polar graph grid and checking your answers to Exercise 9.1.15 of <u>Symmetry, Shape, and Space</u>.

On an empty sketch, choose **Graph: Grid Form: Polar Grid**. Notice that moving the point at rectangular coordinates $(1, 0)$ and polar coordinates $(1, 0°)$ resizes the grid. Making it too big or too small will add or delete rings by changing the scale. Now choose **Graph: Plot Points**, and note that **Polar** (r, θ) is marked. Put in the desired values and click **Plot**. Click **Done** when you have finished putting in your points.

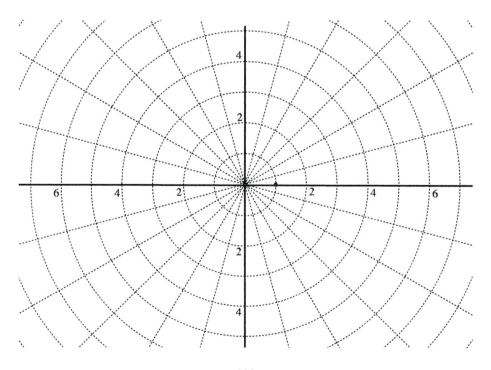

▷ **Exercise 6.** **[SSS 9.1.15]** Using **Graph: Grid Form: Polar Grid** and **Graph: Plot Points**, plot the following points:

(a) $(1, 45°)$

(b) $(2, 90°)$

(c) $(3, 60°)$

(d) $(1, 180°)$

(e) $(1, -90°)$

(f) $(2, 540°)$

Now we will give equations to the software and let it draw the spirals for us. You may recall from Symmetry, Shape, and Space that the equation for the Archimedean spiral is $r = k\theta = \frac{k\pi}{180°}\Theta$ where θ is measured in radians and Θ is measured in degrees. The only result of changing from one type of unit to the other in the pictures is the scale. Below, we have drawn $r = \theta = \frac{\pi}{180°}\Theta$ where one rotation is $2\pi \approx 6.28$, and then $r = \Theta$ where one rotation is 360°. [★ Spirals.gsp: Logarithmic spiral 1 and 2]

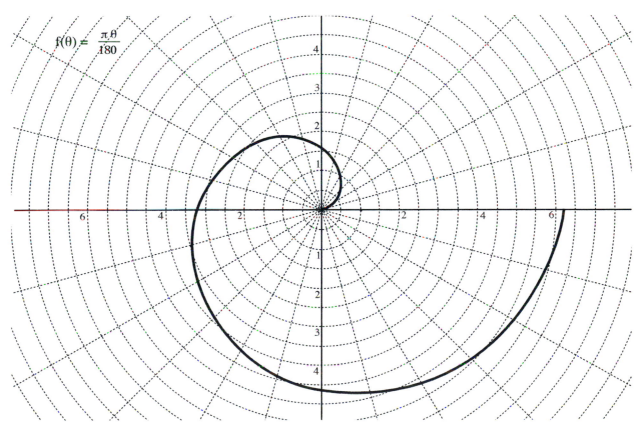

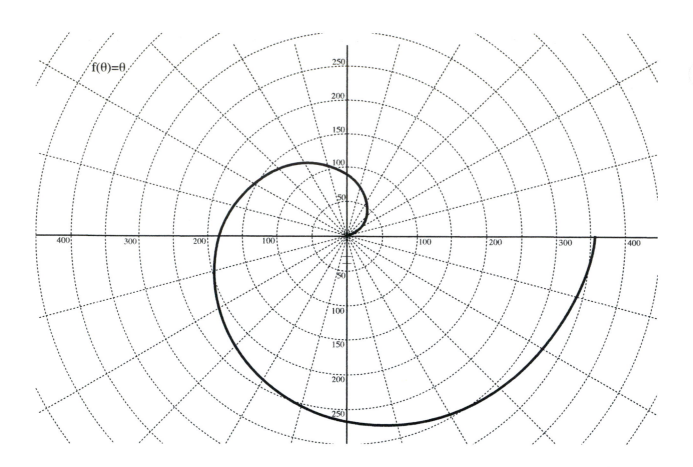

Demonstration: Drawing Archimedean Spirals

1. To graph these pictures, choose **Graph: Grid Form: Polar Grid**.

2. Choose **Graph: New Function**. The default equation in polar coordinates is $r = f(\theta)$ with the angle measured in degrees. (If θ is not an option on your pop-up window but r is there instead, click on the **Equation** button and select $r = f(\theta)$ instead of $\theta = f(r)$.)

3. For the first picture above, we entered $\pi * \theta \div 180$, which is printed on the sketch as $f(\theta) = \frac{\pi \cdot \theta}{180}$. For the second picture, we only had to enter θ to get $f(\theta) = \theta$.

4. Once you have created your function, select it and choose **Graph: Plot Function**. The default window shows the first graph quite well. It does not show the second graph nearly as well. You will need to move the point $(1,0)$ much closer to the origin to resize the picture until you can see the whole of the second curve.

5. You can change the domain to see more than one rotation of either spiral by selecting the curve and choosing **Edit: Properties: Plot: Domain**. Below we have redrawn the graph of $f(\theta) = \frac{\pi \cdot \theta}{180}$ with the domain changed to 0° to 3600° and with the scale adjusted so the whole plot is visible. [★ Spirals.gsp: Logarithmic spiral 3]

112

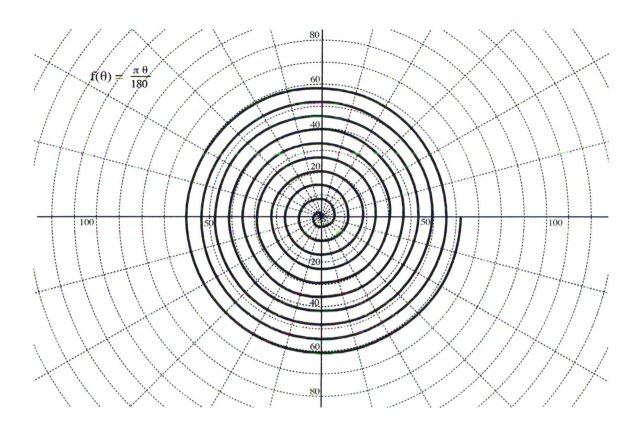

$$f(\theta) = \frac{\pi\,\theta}{180}$$

▷ **Exercise 7.** **[SSS 9.1.16]** Draw the Archimedean spiral $f(\theta) = \frac{\pi}{360}\theta$.

▷ **Exercise 8.** Draw the hyperbolic spiral $f(\theta) = \frac{180}{\pi\theta}$.

▷ **Exercise 9.** Draw the Fermat spiral $\theta = f(r) = \frac{180}{\pi}r^2$. Note that θ is a function of r for this equation. You will need to click on the **Equation** button of the **New Function** box, and select $\theta = f(r)$ instead of $r = f(\theta)$.

▷ **Exercise 10.** Draw the logarithmic spiral $f(\theta) = 2^{\frac{\pi}{180}\theta}$.

113

23. Fibonacci Numbers and Phyllotaxis

Companion to Chapter 9.2 of <u>Symmetry, Shape, and Space</u>

Most of the exercises from the section of <u>Symmetry, Shape, and Space</u> on Fibonacci numbers and phyllotaxis are dedicated to examples from nature. They ask you to analyze actual flowers, pinecones, tree branches, and such. However, a few of the exercises require plotting points on polar graph paper and measuring distances on the sketches. *The Geometer's Sketchpad* facilitates both of these activities.

Open a new sketch and choose the **Graph: Grid Form: Polar Grid** option. You should get a sheet of polar graph paper with points at the origin and $(1, 0°)$. You can move the point at $(1, 0°)$ to resize the grid. Note that if you move the point too far, the grid automatically changes scale. Resize the grid so that you have rings showing at least every two units from the origin and as large a radius (about 30) as you can get in the window. You will need this picture for the following demonstration and exercises, so you may want to save a copy.

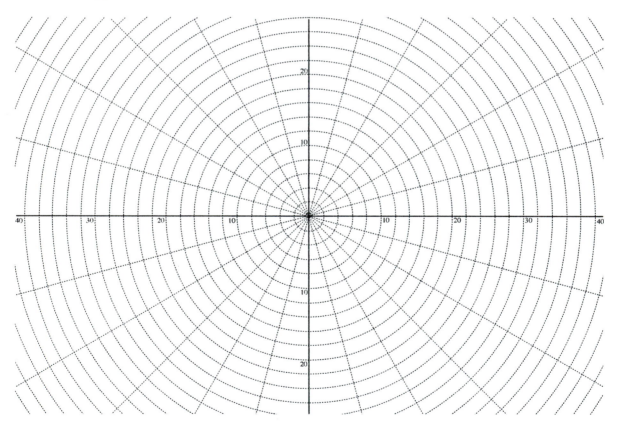

On page 298 of <u>Symmetry, Shape, and Space</u>, you are asked to graph several possible distributions of plant primordia emerging at different angles. There are several ways to do this on a polar graph using *The Geometer's Sketchpad*. One way is to put in each point individually. Open a sheet of polar graph paper. Then choose **Graph: Plot Points**. A window pops up already in polar coordinates with degrees as the units on the angle. Putting in each r and θ value and clicking **Plot** will allow you to plot the points one at a time. You can even put calculations in the windows.

114

The first plot of primordia in <u>Symmetry, Shape, and Space</u> is of florets emerging at a 60° angle. Plotting the first few points individually, we input $(1, 60°)$, $(2, 120°)$, $(3, 180°)$, but then quickly switch to $(4, 4 * 60°)$, $(5, 5 * 60°)$, and so on, letting the computer do the multiplication. Even so, this method is very time-consuming and not very informative once you understand how to plot a single point. Hence, we let the computer do all the calculating and plotting.

Demonstration: Plotting Emerging Primordia (*Plot 30 primordia emerging at an angle of* 60°.)

1. Open a sheet of polar graph paper sized as described above.

2. Choose **Graph: New Function**. In the pop-up box, click on **Equation** and choose $\theta = f(r)$.

3. On the calculator, choose **60**. Then click on **Units** and choose **Degrees**.

4. Now click on *, then r, and then **OK**. You should see $f(r) = 60° r$ in a box on your graph.

5. Choose **Graph: Plot Function**. This gives an Archimedean spiral and is a continuous function.

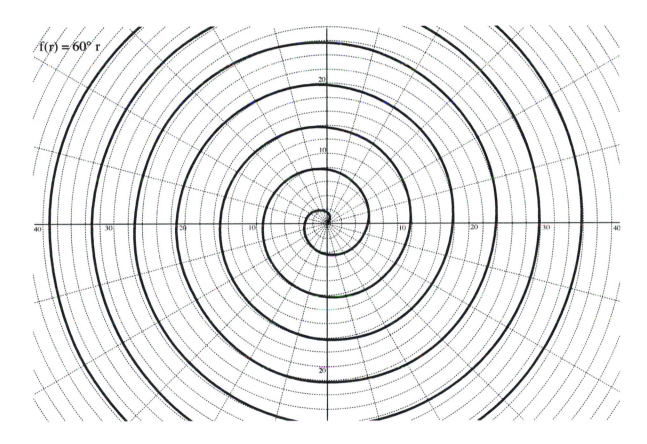

6. We only want the points where r is a whole number between 1 and 30. Change the domain and type of function by choosing **Edit: Properties** and choosing the **Plot** tab in the window. (Note that the function must be highlighted to get the **Plot** option.) Click **Discrete** instead of continuous.

7. If you click **OK** at this stage, you get a graph with 500 points (the default sample size) at unknown (to us) values of r. (Actually these points are equally spaced through the domain chosen.) However, you can see a spiral pattern as shown below:

115

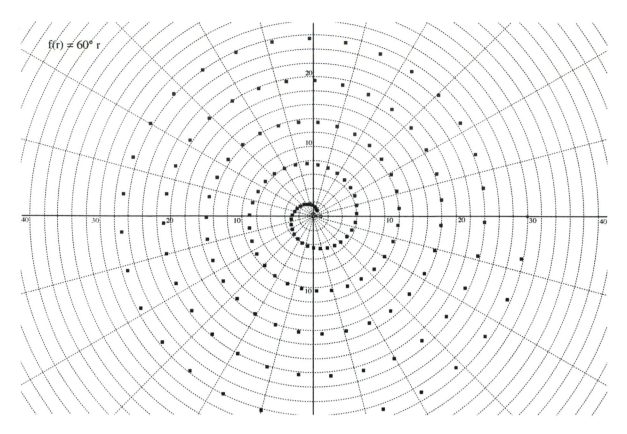

f(r) = 60° r

8. Again, choose **Edit: Properties: Plot**. For the domain, choose $1 \leq r \leq 30$. For **Number of Samples**, select 30 samples and then click **OK**. Now the graph should look like the picture below, modeled on the illustration on page 298 of <u>Symmetry, Shape, and Space</u>.

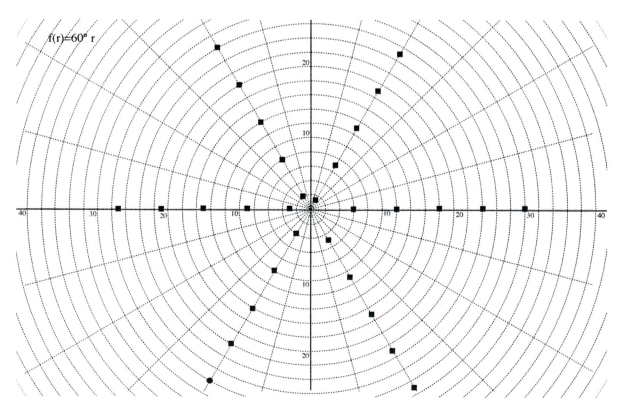

f(r) = 60° r

Following the outline of this demonstration, plot the graphs for the following exercises from Symmetry, Shape, and Space.

▷ **Exercise 1.** **[SSS 9.2.15]** Plot 30 primordia emerging at an angle of $0.5(360°) = 180°$.

▷ **Exercise 2.** **[SSS 9.2.16]** Plot 30 primordia emerging at an angle of $0.6(360°) = 216° = -144°$.

▷ **Exercise 3.** **[SSS 9.2.17]** Plot 30 primordia emerging at an angle of $0.7(360°) = 252° = -108°$.

One of the interesting features of plant growth that is not addressed in the above exercises is that the distances between emerging rings are not generally equal. They tend to follow a Fermat or parabolic spiral, but even that depends on things like rainfall and growing conditions. We will assume equal distance between rings to make the following exercises easier, but even with this assumption, there are some interesting twists. Consider the exercise from Symmetry, Shape, and Space restated below with the pictures redrawn using *The Geometer's Sketchpad*.

▷ **Exercise 4.** **[SSS 9.2.14]** Below are 5 pictures of a growth tip, showing the first primordia. In the first picture primordium #1 has just emerged.

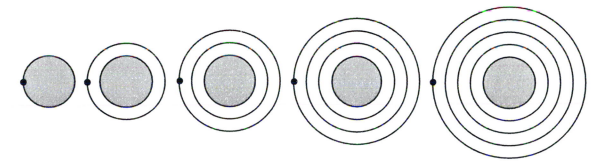

(a) In the second picture, primordium #1 has moved away from the rim. Add primordium #2, placing it on the rim of the shaded disk as far as possible from #1.

(b) In the third picture, draw primordium #2 on the middle ring as positioned in (a). Add primordium #3, placing it on the rim of the shaded disk as far as possible from #1 and #2.

(c) In the fourth picture, draw primordia #2 and #3 on the second and third rings as positioned in (b). Add primordium #4, placing it on the rim of the shaded disk as far as possible from the others.

(d) In the last picture, draw primordia #2, #3, and #4 on successive rings as positioned in (c). Add primordium #5, placing it on the rim of the shaded disk as far as possible from the others.

(e) Will primordium #5 ever recover from its poor start in life?

We have found that when students place the next three primordia on this picture, they typically place them at the east, north, and south points of the circles. We will use the **Measure** and **Calculate** features to show that these placements are not optimal. In *The Geometer's Sketchpad*, it is easier to work with the final sketch rather than five successive sketches.

Demonstration: Placing Primordia [★ Primordia.gsp]

1. Open a sketch and choose **Graph: Grid Form: Square Grid**. Move the point at $(1, 0°)$ if you want to resize your grid. We will be working with circles up to a radius of 6.

2. Choose **Graph: Plot Point**, and plot points at positions $(2, 0°)$, $(3, 0°)$, $(4, 0°)$, $(5, 0°)$, and $(6, 0°)$.

3. Using **Construct: Circle by Center + Point**, construct the circles centered at the origin and through each of the points plotted in Step 2.

4. Select the smallest circle and choose **Construct: Circle Interior**. Hide all of the points shown.

5. Plot a point at position $(-6, 0°)$. Label this point A. This represents the first and oldest primordium.

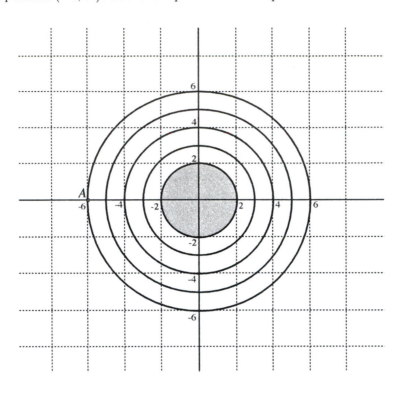

6. Select the second circle (the one of radius 5), and use **Construct: Point on Circle** or the **Point** tool to place point B on this circle. This represents the second primordium.

7. Similarly, place point C on the circle of radius 4, D on the circle of radius 3, and E on the innermost circle of radius 2.

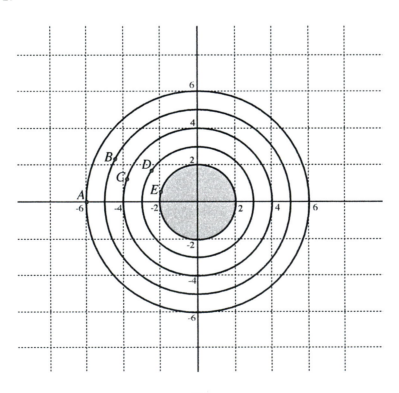

▷ **Exercise 5.** Select points A and B, and choose **Measure: Distance**. A box should appear showing the distance between these two points. Move point B until this distance is as large as you can make it. You should not be surprised that the distance between the points is maximized when the second point is at $(5, 0)$.

▷ **Exercise 6.** Leaving points A and B as placed in Exercise 5, measure the distances from C to A and from C to B. Now choose **Measure: Calculate** to compute the sum of these distances. Move point C until this sum of the distances is as large as you can make it. It may help to enlarge your picture and to change the precision of each measurement to thousandths or ten thousandths using the **Edit: Properties: Precision** command. Once you have the distance maximized, choose **Measure: Coordinates** to get the position of point C. It should NOT be on the y-axis, but it will be very close.

▷ **Exercise 7.** Leaving points A, B, and C as placed in Exercises 5 and 6, measure the distances from D to A, from D to B, and from D to C. Now choose **Measure: Calculate** to compute the sum of these distances. Move point D until this sum of the distances is as large as you can make it. Once you have the distance maximized, choose **Measure: Coordinates** to get the position of point D.

▷ **Exercise 8.** Leaving points A, B, C, and D as placed in Exercises 5, 6, and 7, measure the distances from E to A, from E to B, from E to C, and from E to D. Now choose **Measure: Calculate** to compute the sum of these distances. Move point E until this sum of the distances is as large as you can make it. Once you have the distance maximized, choose **Measure: Coordinates** to get the position of point E.

24. Perspective

Companion to Chapter 10.1 of <u>Symmetry, Shape, and Space</u>

The Geometer's Sketchpad handles the technical aspects of drawing in perspective very well, since these rely basically on ruler and compass techniques. You should read Chapter 10.1 in <u>Symmetry, Shape, and Space</u> first to understand the relationships between the picture plane and the object plane, and the concept of *rabattement*. We begin by drawing two perspective views of a simple set of train tracks, which will help you understand the treatment of parallel lines in perspective.

Demonstration: Perspective View of Centered Train Tracks [Perspective.gsp: Train tracks: centered]

We wish to draw a perspective view of a set of train tracks as viewed by someone standing at the center of the track. First, we construct the train tracks themselves in the object plane, and then the perspective view in the picture plane.

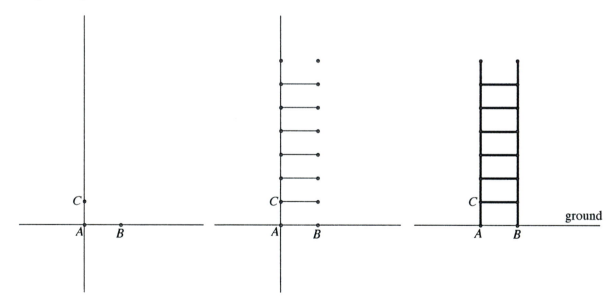

1. Draw a horizontal line, which will represent the ground line. Drag the two points A and B that determine this line a comfortable distance apart to represent the width of the track.
2. Erect a perpendicular to AB through A.
3. Place a point C on this line. The distance between A and C will represent the spacing between railway ties (the crossbars of the track).
4. Draw the line segment AB.
5. Select first point A and then C, and choose **Transform: Mark Vector**.
6. Translate the line segment AB and the points A and B in this direction a number of times to get the length of track you wish to draw in perspective. The last time, translate only the endpoints and not the segment itself.

7. Draw parallel line segments from A and B to the topmost pair of translated points. Use **Display: Line Width: Thick** to thicken these line segments and the crossties. Hide the perpendicular line. These are your train tracks.

8. Draw a ray from A to the first point on the tracks directly above B. Note that we have used **Display: Line Width: Dashed** for the diagonal line to make the drawing less confusing.

9. Select first point B and then A, and choose **Transform: Mark Vector**. Select your diagonal line and point A, and use **Transform: Translate** to translate the diagonal and the endpoint. Repeat to get a diagonal line through each of the points along the left edge of the tracks.

10. Draw a ray from B to the first point of the tracks directly above A and translate this ray and its endpoint in the AB direction. Repeat to get a diagonal line through each of the points along the right edge of the tracks. These diagonal lines will help determine the vanishing points and the spacing of the ties in the perspective view.

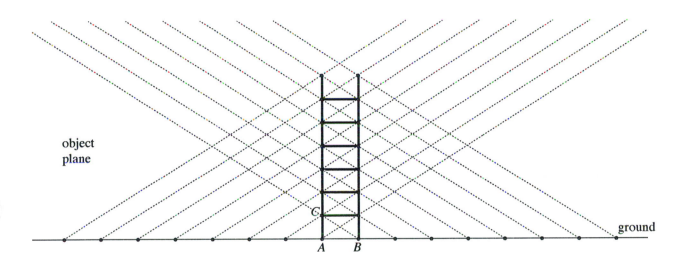

11. Now select the line segment AB (not the line), and use **Construct: Midpoint** to form its midpoint M.

12. Select first A and then M, and choose **Transform: Mark Vector**. Select the diagonal ray through A, and translate to point M.

13. Now select first B and then M, and choose **Transform: Mark Vector**. Select the diagonal ray through B, and translate to point M. You now have an extra pair of diagonal lines centered on the tracks.

14. Select M and the line AB, and choose **Construct: Perpendicular Line**. Use the **Point** tool to place two points on this line, M' and V, as shown on the following page. (You may have to resize the window, or shrink the object plane drawing by dragging C or the ground line.)

15. Select first M and then M', and choose **Transform: Mark Vector**. Select the line AB (not the line segment), the points A and B, the endpoints of the diagonal rays, and the two diagonal rays going through M. Translate these in the MM' direction. The new horizontal line will be the ground line for the perspective drawing.

16. Use **Construct: Parallel Line** to draw a line through V parallel to AB. This line will form the horizon for the perspective drawing, and V will be the principal vanishing point.

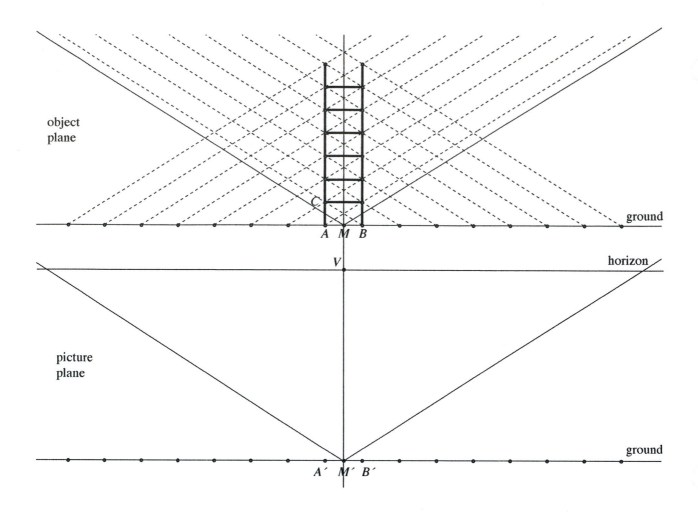

17. Draw the line segments from A' and B' (the translates of A and B) to V.

18. Construct the intersections of the two rays going through M' with the horizontal line through V, and label these V_1 and V_2. (Select the first point and choose **Display: Show Label**. In the blank for the label type **V[1]**. Make sure the option **Show Label** is checked. Repeat for V_2.)

19. Starting at B' and working to the right, draw dashed line segments connecting V_1 to each of the points on $A'B'$ to the right of M'.

20. Starting at A' and working to the left, draw dashed line segments connecting V_2 to each of the points on $A'B'$ to the left of M'. The intersections of these two families of lines will determine the spacing of the ties in the perspective drawing. (For convenience, below we show only the picture plane.)

122

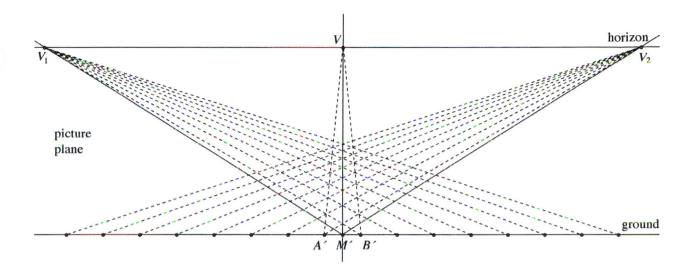

21. Place points along the line segments $A'V$ and $B'V$ where the segments intersect the diamonds formed by the crossing diagonal lines.

22. Draw a thick line segment along $A'V$ connecting A' to the topmost of the points created in Step 21. Repeat along $B'V$. Draw horizontal line segments connecting corresponding points on $A'V$ and $B'V$. (Again we show only the picture plane.)

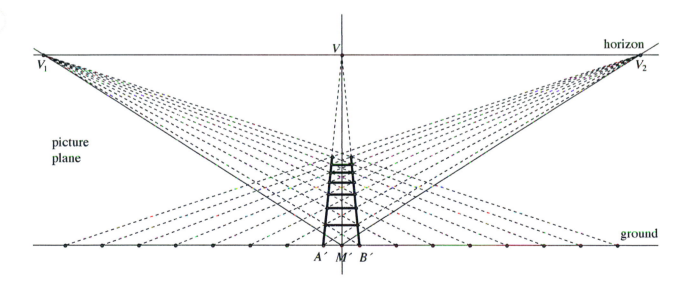

23. To make the drawing less confusing, use **Display: Hide Objects** to hide all the diagonal lines, the line segments $A'V$ and $B'V$, the line MM', the points M, M', V_1, and V_2, and the strings of points along the two horizontal ground lines.

24. The points A, C, and V should be draggable. These can change, respectively, the width of the track, the spacing between crossties, and the vanishing point of the perspective drawing.

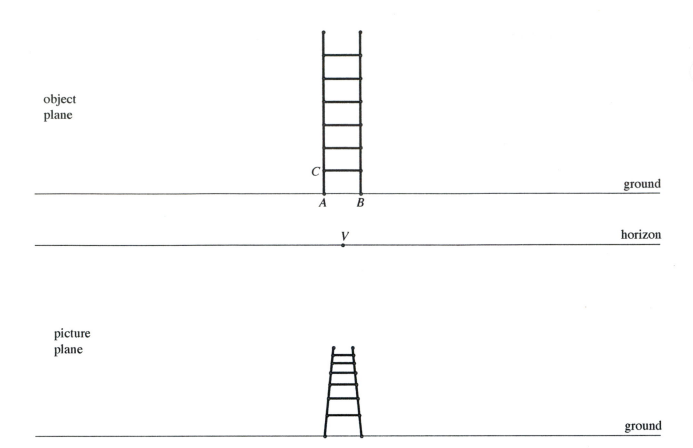

object
plane

C

ground

A *B*

V

horizon

picture
plane

ground

▷ **Exercise 1.** Draw a centered perspective view of a set of train tracks as in the previous demonstration.

Demonstration: Perspective View of Skew Train Tracks [Perspective.gsp: Train tracks: skew]

Next we wish to draw a perspective view of a set of train tracks as viewed by someone standing to the side of the track. First, we will construct the train tracks themselves in the object plane, and then we will construct the perspective view in the picture plane. Many of the steps duplicate those given for the centered view, but there will be a few key differences.

1. Draw a horizontal line DE, which will represent the ground line. Select point D on the line and the line itself, and use **Construct: Perpendicular Line** to draw a perpendicular line. The vanishing point will lie on this line.

2. Place two points A and B on the horizontal line a comfortable distance apart to represent the width of the track. Do not use either of the points D or E.

3. Repeat Steps 2–13 from the previous demonstration to draw train tracks based at A and B, midpoint M, and the families of diagonals.

4. Use the **Point** tool to place two points D' and V on the perpendicular line through D constructed in Step 1.

5. Select first D and then D', and choose **Transform: Mark Vector**. Select the ground line DE, the points A and B, the endpoints of the diagonal rays, and the two diagonal rays going through M. Translate these in the DD' direction. The new horizontal line will be the ground line for the perspective drawing.

6. Use **Construct: Parallel Line** to draw a line through V parallel to DE. This line will form the horizon for the perspective drawing, and V will be the principal vanishing point.

124

7. Repeat Steps 17–20 of the previous demonstration to construct and label V_1 and V_2 and the families of diagonals.

8. Place points where the diagonal lines cross each other at the points corresponding to the intersection of the diagonals and the line AC in the object plane. See the diagram below.

9. Draw lines through the points created in Step 8 parallel to the horizontal line DE.

10. Place points where these horizontal lines cross the line segments $A'V$ and $B'V$.

11. Draw a thick line segment along $A'V$ connecting A' to the topmost of the points created in Step 10. Repeat along $B'V$. Draw horizontal line segments connecting corresponding points on $A'V$ and $B'V$.

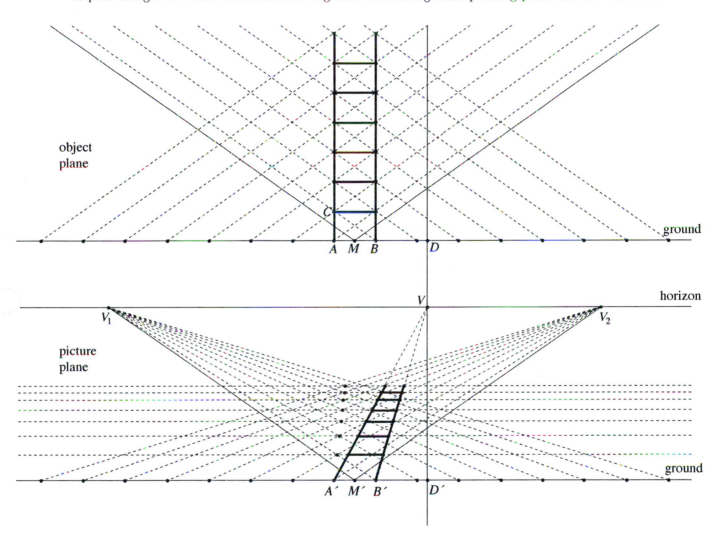

12. To make the drawing less confusing, hide all the diagonal lines, the dashed horizontal lines, the line segments $A'V$ and $B'V$, the line DD', the points D, E, D', M, M', V_1, and V_2, and the strings of points along the two horizontal ground lines.

13. The points A, C, and V should be draggable. These can change, respectively, the width of the track, the spacing between crossties, and the vanishing point of the perspective drawing.

▷ **Exercise 2.** Draw a skew perspective view of a set of train tracks as in the demonstration. Note that dragging points A and B so that V is centered between them should give the centered view of the train tracks.

125

▷ **Exercise 3.** Modify the method of the first demonstration to draw a centered perspective view of the tiled floor shown below. (The only change you will have to make is the addition of more diagonal lines.)

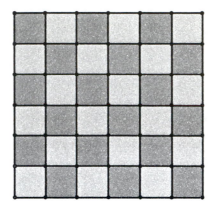

▷ **Exercise 4.** Modify the method of the second demonstration to draw a skew perspective view of the tiled floor shown above. [★ Perspective. gsp: Square tiling]

▷ **Exercise 5.** [SSS 10.1.4] Modify the method of the first demonstration to draw a centered perspective view of the tiled floor shown below.

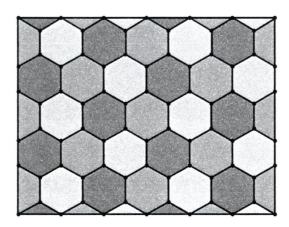

▷ **Exercise 6.** Modify the method of the second demonstration to draw a skew perspective view of the tiled floor shown above. [★ Perspective. gsp: Hexagonal tiling]

One-Point Perspective

Next we will investigate perspective drawings of three-dimensional objects such as rectangular boxes. If you can draw a box in perspective using *The Geometer's Sketchpad*, then theoretically you will be able to draw other objects—cones, houses, Fluffy the Dog, etc.—by approximating them with a series of rectangles. The first method we will demonstrate is the one-point perspective. This method assumes that one face of your box is parallel to the ground line. To simplify the process, we also assume that the height has already been scaled correctly.

Demonstration: One-Point Perspective [★ Perspective.gsp: One point perspective]

We construct first the footprint of the box itself and an indication of its height in the object plane, and then the perspective view in the picture plane.

1. Draw a horizontal line AB representing the ground line in the object plane.

2. Place a point C (anywhere but on the line AB). Construct the line through C parallel to AB, and place a point D on this line.

3. Construct the line through D perpendicular to CD.

4. Construct the line through C perpendicular to CD. Place a point F on this line.

5. Construct the line through F perpendicular to CF (or parallel to CD). Construct the point of intersection E of this line and the line from Step 3.

6. Draw the line segments CD, DE, EF, and FC. Thicken these using **Display: Line Width: Thick**. This rectangle is the footprint of your box. Hide the lines CD, DE, EF, and FC.

7. Construct a line through A and perpendicular to AB. Place three points on this line, and label them Eye, V (for vanishing point), and A'.

8. Draw a line through V parallel to AB. This line will be the horizon for your drawing. Draw another line through A' parallel to AB. This line will be the ground line for the picture plane.

9. Place another point G (anywhere but on any of the lines constructed). Construct a line through G perpendicular to AB. Place another point H on this line. Draw the line segment GH and hide the line. This line segment represents the height of your box.

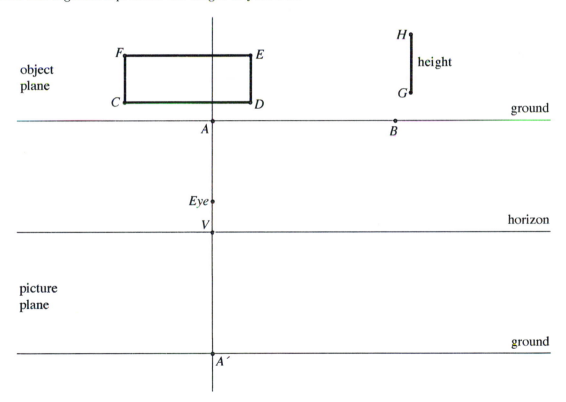

10. Now draw line segments from C, D, E, and F to Eye. Place points C', D', E', and F' where these line segments intersect the ground line AB.

11. Draw lines through C', D', E', and F' perpendicular to AB.

12. Construct the intersections of the lines constructed in Step 11 through C' and D' with the horizontal ground line through A'. Label these points C'' and D''.

13. Draw line segments $C''V$ and $D''V$.

127

14. Construct the intersection F'' of $C''V$ and the vertical line through F'. Similarly, construct the intersection E'' of $D''V$ and the vertical line through E'.

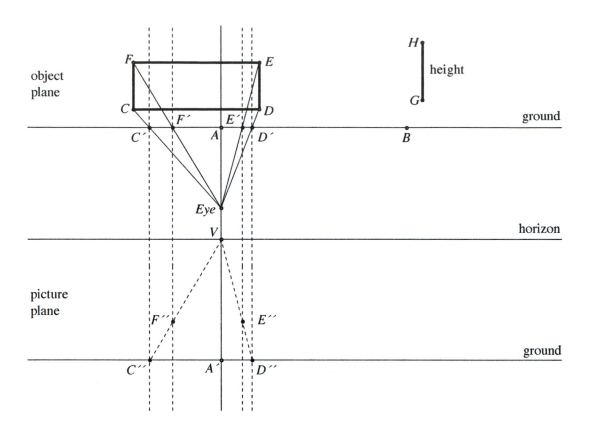

15. Draw the line segments $C''D''$, $D''E''$, $E''F''$, and $F''C''$. Thicken these line segments. These form the perspective view of the footprint of the box.

16. Select first point G and then C''', and choose **Transform: Mark Vector**. Translate the height segment GH into position at C''.

17. Translate another copy of the height segment along vector GD''.

18. Connect the endpoints of these two segments to form the rest of the front face of the box.

19. Draw line segments connecting the upper front corners of the box to the vanishing point V.

20. Place points at the intersections of the line segments drawn in Step 19 with the vertical lines $F'F''$ and $E'E''$.

21. You now have all eight vertices of the box. Connect them with thick line segments to complete the one-point perspective view of the box.

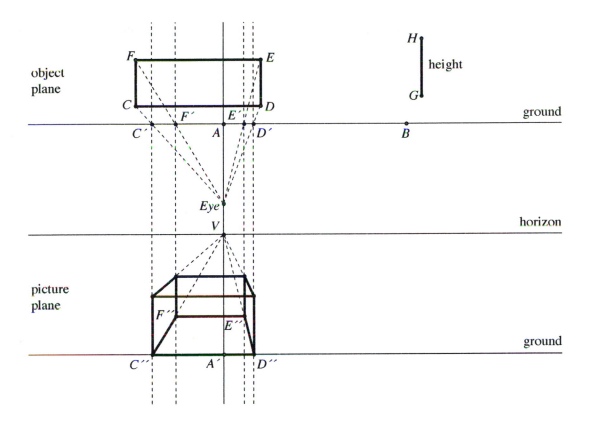

22. To make the drawing less confusing, hide the points A, B, A', C', D', E', and F' and all of the line segments shown as dashed in the picture above. Hide the labels of all points except C, D, F, H, V, and Eye.

23. The points C, D, F, H, V, and Eye should be draggable. These can change, respectively, the position of the box, its width, its depth, its height, the position of the vanishing point of the perspective drawing, and the position of the eye of the viewer.

✳ ▷ **Exercise 7.** [SSS 10.1.8] Draw a one-point perspective of a box as demonstrated above. Experiment with moving the position of the viewer's eye and the vanishing point.

Demonstration: Two-Point Perspective [★ Perspective.gsp: Two point perspective]

We construct first the footprint of the box itself and an indication of its height in the object plane, and then the perspective view in the picture plane. We assume that the box has one vertex on the ground line.

1. Draw a horizontal line AB representing the ground line in the object plane.

2. Place a point C on the line AB. Construct a line segment CD at an angle to AB.

3. Construct the rectangular footprint $CDEF$, the eye, the vanishing point, the point A', and the height of the box as in Steps 3–9 of the previous demonstration.

4. Construct two lines through Eye parallel to the sides of the box. Place points X and Y at the intersections of these lines and the ground line AB.

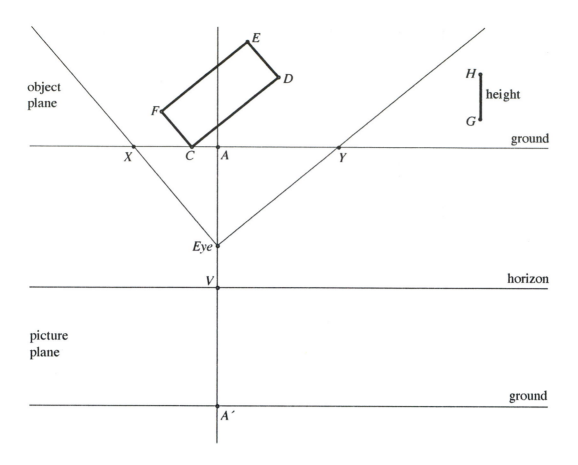

5. Now draw line segments from C, D, E, and F to *Eye*. Place points D', E', and F' where these line segments intersect the ground line AB. Note that $C' = C$.

6. Draw lines through C, D', E', F', X, and Y perpendicular to the ground line AB.

7. Construct the intersection of the line of Step 6 through X and the horizontal line through V, and label this point V_1. Similarly, construct V_2 directly below point Y.

8. Construct and label the point C'' at the intersection of the vertical line through C and the horizontal line through A'.

9. Draw line segments $C''V_1$ and $C''V_2$.

10. Place point D'' at the intersection of $C''V_2$ and the vertical line through D'. Similarly, place F'' at the intersection of $C''V_1$ and the vertical line through F'.

11. Draw the line segments $F''V_2$ and $D''V_1$. Label the intersection of these segments as E''.

12. Draw the line segments $C''D''$, $D''E''$, $E''F''$, and $F''C''$. Thicken these line segments. These form the perspective view of the footprint of the box.

130

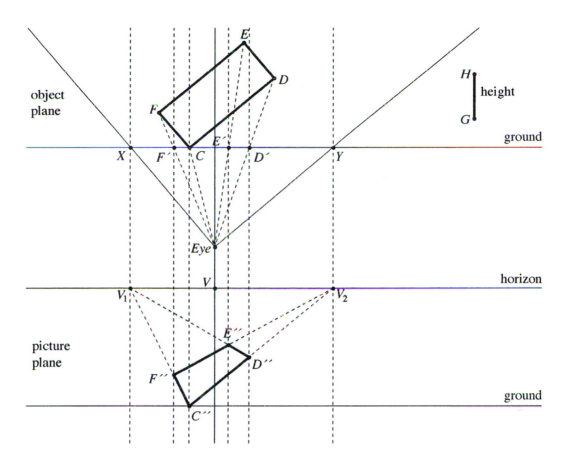

13. Select first point G and then C'', and choose **Transform: Mark Vector**. Translate the height segment GH into position at C''.

14. Connect the endpoint of this vertical segment to points V_1 and V_2.

15. Construct the intersections of the two line segments of Step 14 with the vertical lines $F'F''$ and $D'D''$.

16. Construct the line segments connecting the points of Step 15 with V_2 and V_1, respectively.

17. Construct the intersection of the line segments of Step 16. Note that this intersection should fall on the segment $E'E''$.

18. You now have all eight vertices of the box. Connect them with thick line segments to complete the perspective view of the box.

131

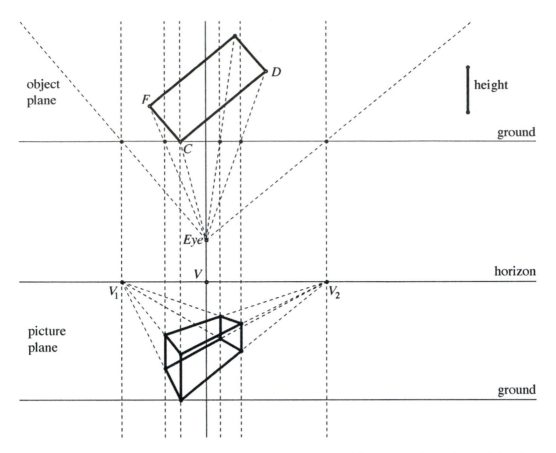

19. To make the drawing less confusing, hide the points A, B, A', D', E', and F' and all of the line segments shown as dashed in the picture above. Hide the labels of all points except C, D, F, H, V, and Eye.

20. The points C, D, F, H, V, and Eye should be draggable. These can change the position of the box, its width, its depth, its height, the position of the vanishing point of the perspective drawing, and the position of the eye of the viewer.

▷ **Exercise 8.**   Draw a two-point perspective of a box as demonstrated above.

▷ **Exercise 9.   [SSS 10.1.9 and 10]** Using your drawing from Exercise 8, experiment with moving the position of the picture plane (move the point A' to do this). Unfortunately, the resulting picture does not accurately represent the height unless we completely redraw the perspective view. What is the effect of moving the picture plane?

▷ **Exercise 10.   [SSS 10.1.11 and 12]** Using your drawing from Exercise 8, experiment with moving the position of the eye. What is the effect of moving the viewpoint or eye?

▷ **Exercise 11.   [SSS 10.1.13 and 14]** Using your drawing from Exercise 8, experiment with moving the position of the principal vanishing point V. What is the effect of moving the horizon?

25. Optical Illusions

Companion to Chapter 10.2 of <u>Symmetry, Shape, and Space</u>

When it comes to optical illusions, even the people who create them can be misled by them. *The Geometer's Sketchpad* gives you the opportunity to create your own illustrations. That way, you know for sure that the line segments are equal, parallel, straight, or whatever aspect of a drawing you may doubt. It is also relatively easy to make adjustments to a drawing if you have used the menu options to create it. In the following exercises, you will create many of the illusions shown in <u>Symmetry, Shape, and Space</u>. If you have the opportunity, have several people play with your figures to see where they think the illusions are maximized, and compare the results.

Demonstration: Müller-Lyer Illusion [★ Optical Illusions.gsp: Muller-Lyer]

First, we construct the classic Müller-Lyer illusion. Drawing two equal line segments is fairly easy. Getting all the arrow parts at the same angle with identical lengths is more complicated. We will use menu commands to create many of these parts.

1. Construct a horizontal line segment CD. Construct the lines through each endpoint and perpendicular to the line segment: Select the segment and each point, and then choose **Construct: Perpendicular Line**.

2. Construct a point A on one of the vertical lines using the **Point** tool. Select this point and the horizontal line segment CD, and choose **Construct: Parallel Line**. This should give you the rectangle $ABDC$ as below. Construct the fourth corner, point B, by selecting the lines that intersect there and choosing **Construct: Intersection**. Use the **Segment** tool to construct AB.

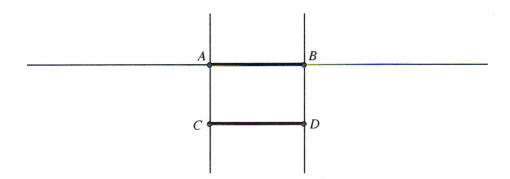

3. Construct a short slanted line segment beginning at A that goes up and to the right.

4. Double-click on line AB to mark it as the mirror line. (It should flash when this occurs.) Then select your slanted segment, and choose **Transform: Reflect**. This should give you the arrow at A.

5. Select A and then D, and choose **Transform: Mark Vector**. Then select the two parts of the arrow, and choose **Transform: Translate**. This should give you the arrow pointing in at D.

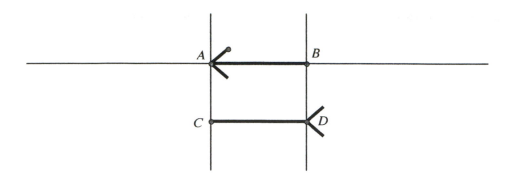

6. Double-click on AC to mark it as the mirror line. Select the arrow at A, and use **Transform: Reflect** to reflect it across the line.

7. Select A and then C, and choose **Transform: Mark Vector**. Select the reflected arrow from Step 6, and translate it to C.

8. Select A and then B, and choose **Transform: Mark Vector**. Select the reflected arrow from Step 6, and translate it to B.

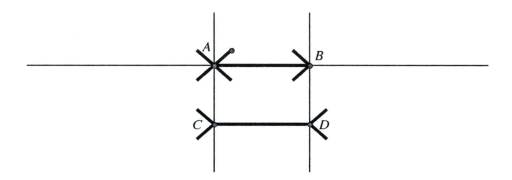

9. Hide all of your construction lines and vertices except the vertices A, B, C, and D along with the vertex at the end of your first arrow piece.

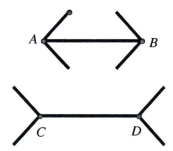

Note that because we used the **Construct** and **Transform** commands to create the figure, the relationship between relevant parts is maintained when you move the vertices. Dragging point D will elongate the horizontal line segments. Dragging the vertex at the end of the arrowhead will change the length and angle of the arrows.

▷ **Exercise 1.** Construct the Müller-Lyer illusion as described above. Change the length of the segment CD, the angle and length of the arrow parts, and the distance between the line segments by moving the key points. Determine where you think the illusion is most effective, and compare your drawing with those of others doing the same exercise.

134

▷ **Exercise 2.** Draw the two illusions below so that the relationships between relevant parts are maintained when you move the vertices.

The Ponzo illusions can also be created and adjusted easily using *The Geometer's Sketchpad*.

Demonstration: Ponzo Illusion [★ Optical Illusions.gsp: Ponzo]

1. Construct a horizontal line segment. From the left endpoint on that line, construct a slanted line segment that goes up and to the right. Reflect this segment across the horizontal line.

2. Construct a circle inside this angle and centered at point A on the horizontal line.

3. Place a point B on the horizontal line, and mark AB as the translation vector. Translate a copy of the circle centered at A using the **Transform: Translate** command. If the two circles intersect, move B along the line until they do not.

4. Hide the horizontal line and any stray points. Dragging A and B will change the position of the circles. Dragging the radius point on the first circle should change the size of both circles. Dragging the endpoint for the angle will change the size of the angle.

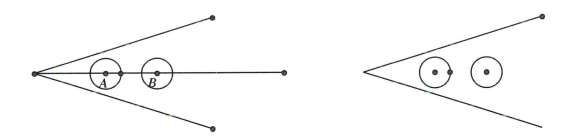

▷ **Exercise 3.** Construct a Ponzo illusion as described above. Change the angle to find where you think the illusion is most effective. Compare your answer with those of others doing the same exercise.

▷ **Exercise 4.** Reflect your Ponzo illusion from Exercise 3 across a line perpendicular to AB so that the angle opens to the left. Hide the original construction and the perpendicular line. Move the key points to maximize the illusion. Compare the result with your result from Exercise 3.

▷ **Exercise 5.** Alter your illusion from Exercise 3 by changing the size and/or number of circles and the distances between them and between the first circle and the angle vertex. What is the effect of making these changes? Maximize the illusion, and compare your answer with those of others doing the same exercise.

▷ **Exercise 6.** Construct a set of Titchener circles. Using the **Transform: Translate** command will ensure that the two inner circles are congruent. Using the **Transform: Rotate** command (with an angle of 60°) to construct the outer rings of circles will ensure they remain congruent when you adjust the size. [★ Optical Illusions.gsp: Titchener]

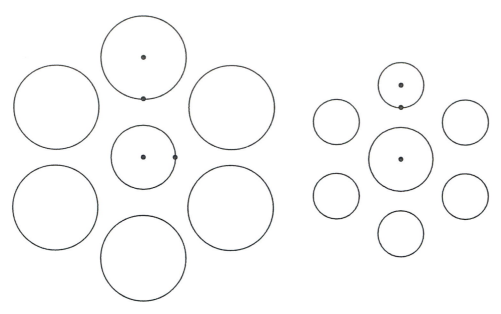

▷ **Exercise 7.** [SSS 10.2.10] Using your sketch from Exercise 6, explore the effect of increasing and decreasing both the size of the surrounding circles and the distance of the surrounding circles from the center circle.

Demonstration: Zöllner Illusion [★ Optical Illusions.gsp: Zollner]

We will next construct the original Zöllner illusion. We will use *The Geometer's Sketchpad* to draw a set of parallel line segments and then ask that you add slashes to these so that the lines appear to converge. Then we will create a dynamic sketch so that the length and angle of the slashes can be varied.

1. Construct a horizontal line *AB*. Select point *A* and the line, and choose **Construct: Perpendicular Line**. Construct a line segment along the perpendicular line, and then hide the perpendicular line.

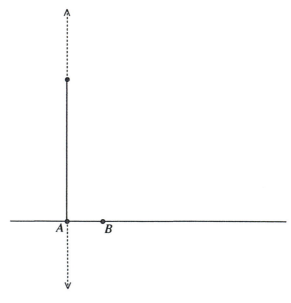

2. Select the two points A and B in order, and choose **Transform: Mark Vector**.
3. Select the vertical line segment, and choose **Transform: Translate** seven times so that you have eight parallel lines.
4. Hide the horizontal line and any stray points. When working by hand, this is where we start Zöllner's illusion. Print a few copies of the sketch, and try adding lines by hand to make the illusion work before continuing.

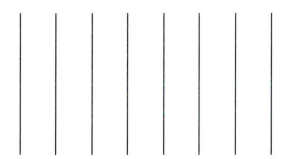

5. Now open a new sketch to draw a dynamic version of Zöllner's illusion. Draw a horizontal line AB and a line through A perpendicular to the horizontal line.
6. Place a point C on the vertical line a little above A and another point D a bit off the line.
7. Select first point D and then C, and choose **Transform: Mark Vector**. Translate point C using this vector.
8. Draw the line segment connecting D to the translated point constructed in Step 7. This is your first diagonal slash mark.
9. Select first point A and then C, and choose **Transform: Mark Vector**. Translate the short slanted segment and its central point in this direction several times.
10. Now translate the last point on the vertical line (not the line segment) in the chosen direction. Construct the line segment connecting this top point to A.

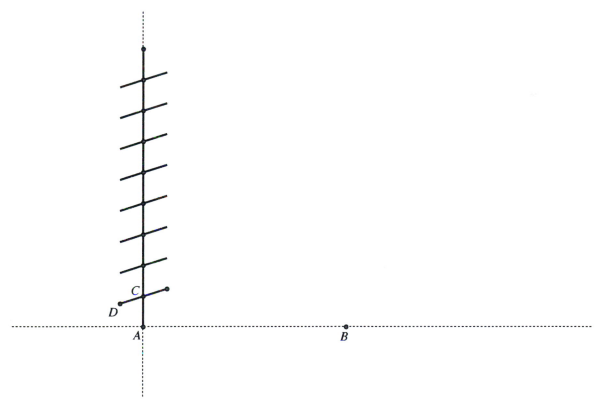

11. Construct the line segment AB, and use **Construct: Midpoint** to find its midpoint E.

12. Construct the line segment AE, and use **Construct: Midpoint** to find its midpoint F.

13. Construct the line through F and perpendicular to AB. Select this line and choose **Transform: Mark Mirror**.

14. Select the vertical segment through A and all of its slanted slashes, and use **Transform: Reflect** to get the reflection.

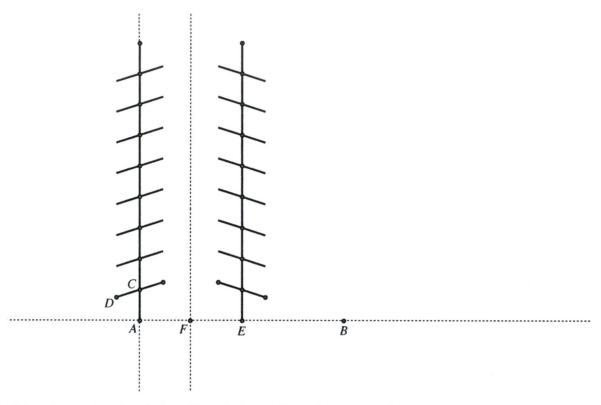

15. Select first point A and then B, and choose **Transform: Mark Vector**. Translate the two vertical line segments and their slashes in this direction three times.

16. Hide the long lines and all of the points except A, B, C, and D.

17. Dragging point B will change the spacing of the vertical segments. Dragging point C will change the spacing of the diagonal line segments. Dragging point D will change the angle and length of these slashes.

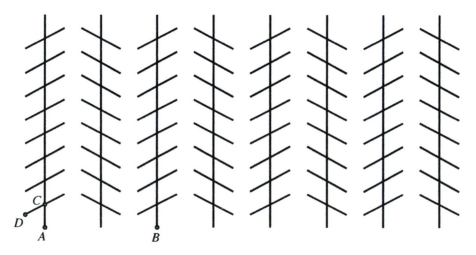

▷ **Exercise 8.** Construct Zöllner's illusion as in the previous demonstration. Move the vertices to change the slope of the slash marks, the distance between the slash marks, and the distance between the parallel lines. Compare the picture that you feel maximizes the illusion with those of others working on the same exercise. And if you find yourself doubting that the vertical lines are parallel, you can always hide the slash marks or make them horizontal and see for yourself that they really are.

▷ **Exercise 9.** [SSS 10.2.12] Reverse Zöllner's illusion by constructing a horizontal line and a line segment not quite perpendicular to your original line. Modify the procedure above to add equally spaced slashed marks to the nonperpendicular line segment. Reflect the line segment and its slashes across a line perpendicular to the horizontal line, and translate this pair to get four copies (as shown below without the slash marks). Use the basic method of the demonstration for Zöllner's illusion, so that moving a few key points changes the spacing and angle of the slashes. Adjust your slashes until the nonparallel lines appear parallel.

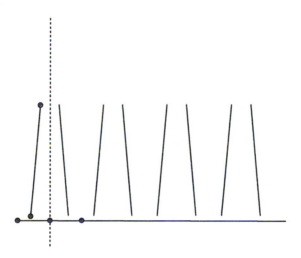

▷ **Exercise 10.** [SSS 10.2.13] In this exercise we make use of the ideas from the figures below to try to reverse the illusions.

(a) Draw a square and a field of circles so that it appears that the sides of the square bow out.

(b) Draw a pair of parallel lines and a field of other lines to make the parallel lines appear to bend inward.

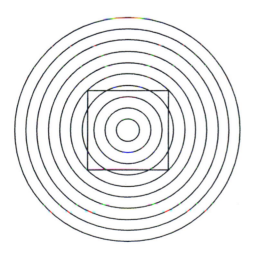
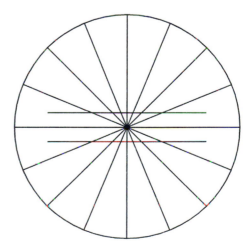

▷ **Exercise 11.** Use *The Geometer's Sketchpad* to draw the following impossible objects. Use the **Construct** and **Transform** menus to make dynamic drawings of the objects, so that moving a few key points changes the shape but preserves the illusion.

 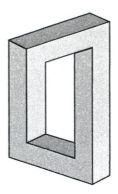

▷ **Exercise 12.** [SSS 10.2.22] Create your own drawing of an "impossible object," different from those shown above.

26. Noneuclidean Geometry

Companion to Chapter 11.1 of <u>Symmetry, Shape, and Space</u>

The Geometer's Sketchpad provides excellent tools for the investigation of the Poincaré disk model of noneuclidean geometry, which would otherwise be beyond the skills assumed for this text. Look in the folder that came with the software, and you will find the sketch **Sketchpad: Samples: Sketches: Investigations: Poincare Disk.gsp**. This file provides a basic configuration and a set of custom tools that allow one to draw figures in hyperbolic geometry. However, remember that the Poincaré disk is essentially a map of hyperbolic space, with all the defects commonly found in maps of our earth. The infinite hyperbolic plane itself is represented by the interior of the disk. In the sketch **Poincare Disk.gsp**, one point is shown on the rim of the circle and labelled **Radius**. You can move this point if, for any reason, you want to change the radius of the universe. Lines are represented as arcs from circles that meet the boundary of the hyperbolic world at right angles. Line segments are represented by smaller arcs of these circles. Distance is badly distorted, especially as one moves toward the edge of the world. However, this model does represent angles correctly which is its great advantage over other possible maps.

General Instructions: Open the document **Sketchpad: Samples: Sketches: Investigations: Poincare Disk.gsp** that came with your copy of *The Geometer's Sketchpad*. You should see the following screen:

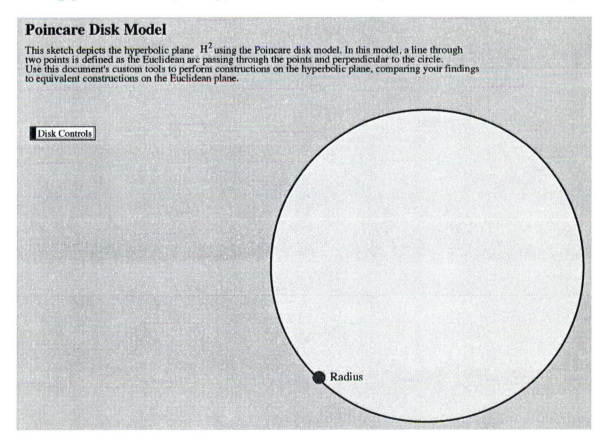

141

Clicking on the **Disk Controls** button shows the center of the Poincaré disk and the point on the radius that defines this circle. Moving the center point slides the disk around, and moving the radius point increases or decreases the size of the circle. Press the **Custom Tool** button (at the bottom of the tool bar), and you will see the following list of tools:

Hyperbolic Segment: Given two points inside the disk, this draws the hyperbolic line segment connecting them.

Hyperbolic Line: Given two points inside the disk, this draws the hyperbolic line connecting them.

Hyperbolic P. Bisector: Given two points inside the disk, this draws the hyperbolic perpendicular bisector of the hyperbolic line segment between them.

Hyperbolic Perpendicular: Given three points inside the disk, this drops the hyperbolic perpendicular from the first point to the hyperbolic line connecting the second and third points.

Hyperbolic A. Bisector: Given three points inside the disk, this draws the hyperbolic line that bisects the hyperbolic angle formed by the three points.

Hyperbolic Circle by CP: Given two points inside the disk, this draws a hyperbolic circle centered at the first point and passing through the second.

Hyperbolic Circle by CR: Given three points inside the disk, this draws a hyperbolic circle centered at the first point with radius equal to the hyperbolic distance between the second and third points.

Hyperbolic Angle: Given three points inside the disk, this measures the hyperbolic angle they form.

Hyperbolic Distance: Given two points inside the disk, this measures the hyperbolic distance between them.

We will use these tools to replace all of the tool bar and **Construct** menu commands. These custom tools will be available as long as **Poincare Disk.gsp** is open in the background. The only old tools and commands you are allowed to use are the **Point** tool and the commands **Construct: Point on Object** and **Construct: Intersection**; when using these, you must make sure that all points fall within the disk.

We will illustrate the use of some of these new tools and then give you some exercises.

Demonstration: The Sum of the Angles in a Hyperbolic Triangle [★ Hyperbolic Geometry.gsp: Hyperbolic triangle]

1. Open up the sketch **Sketchpad: Samples: Sketches: Investigations: Poincare Disk.gsp**. With this sketch open, choose **File: New Sketch** to open a new drawing window.

2. Press the **Custom Tool** button and slide to the right. You will see an option labelled **Poincare Disk** with an arrow. Slide down to **Poincare Disk**; to the right of that you will see the list of hyperbolic tools. These are available as long as **Poincare Disk.gsp** is open in the background. To use the **Hyperbolic Segment** tool, click this option on the menu. The **Custom Tool** option will be highlighted on the tool bar, and the selection arrow will show a point. Click on the drawing screen once, and a point will appear labelled **P. Disk Center**. Place another point, which will be labelled **P. Disk Radius**, and the disk will automatically be drawn. Click two more places within this disk, and the hyperbolic line segment between these points will be generated. Label the two endpoints A and B.

3. To draw a second hyperbolic line segment, make sure that **Custom Tool: Poincare Disk: Hyperbolic Segment** is still chosen. Place the **Selection Arrow** over point A and click (it should light up), and then place another point. The line segment from A to the new point should appear. Label this point C. Note that you do not have to select **P. Disk Center** or **P. Disk Radius** since these will be the same as in Step 2.

4. To draw the third line segment, make sure that the **Custom Tool** option is highlighted and click on points B and C. You now have a hyperbolic triangle.

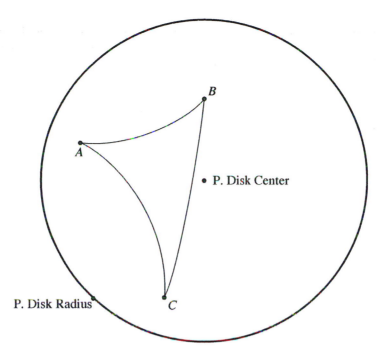

5. Now we want to measure the angles of this triangle. Hold down the **Custom Tool** button and slide to the right, down to **Poincare Disk**, and again to the right to choose **Hyperbolic Angle**.

6. Select in order points A, B, and C. The hyperbolic angle measure for $\angle ABC$ should appear.

7. Repeat to measure the hyperbolic angles $\angle BCA$ and $\angle CAB$.

8. Choose **Measure: Calculate**. While the calculator window is open, click on the first angle measurement in the drawing window, then click $+$, then the second angle measure and another $+$, and last the third angle measure. Finally, click **OK**, and the sum of the angles should appear on the drawing screen.

$$m\angle ABC = 40.1°$$
$$m\angle BCA = 12.9°$$
$$m\angle CAB = 25.9°$$

$$m_1 + m_2 + m_3 = 78.89°$$

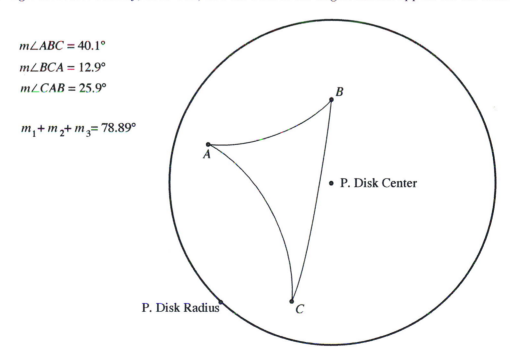

An alternate (rather longer, but informative) way to use the hyperbolic custom tools is to choose the option **Custom Tool: Show Script**. For example, if **Custom Tool: Poincare Disk: Hyperbolic**

Segment is chosen, a window full of text labelled **Hyperbolic Segment Script** will appear at the right of your screen. Move this if necessary so it doesn't obscure the drawing window. At the top of the script window, you will see the following:

> **Assuming:**
> 1. Point P. Disk Center
> 2. Point P. Disk Radius
> **Given:**
> 1. Point A
> 2. Point B

and then a number of steps. Select the center of the disk; the first line in the script window should be highlighted. Then select, in order, the radius point, the first point A you placed, and then the second point B. All of these labels should now be highlighted in the **Hyperbolic Segment Script** window. At the bottom of this window, two buttons should now appear labelled **Next Step** and **All Steps**. Clicking the first button repeatedly will walk you through the process of constructing a hyperbolic line segment step by step, while the second button will quickly perform all of the necessary steps at once. Use either of these to construct the hyperbolic line segment from A to B. It is very interesting to see how many steps are involved and how much is hidden in the final drawing, so we suggest that you try this method at least once. Also, if merely clicking on a tool does not give the desired results, consulting the script can tell you what is going wrong.

In any case, you now have a hyperbolic triangle with each angle measured and the sum of these angles computed. Note that the angle sum is significantly smaller than 180°. Play with your triangle by moving the vertices.

▷ **Exercise 1.** Construct a hyperbolic triangle $\triangle ABC$ as above, measure each of the hyperbolic angles, and find their sum. Move the points A, B, and C around. What is the smallest angle sum you can get? What is the largest angle sum?

Demonstration: Parallel Lines [★ Hyperbolic Geometry.gsp: Parallel lines]

Recall that the definition of parallel lines is that they are lines that never intersect. Other properties follow from this, but many of these facts are only true in Euclidean geometry, such as the fact that parallel lines are always the same distance apart.

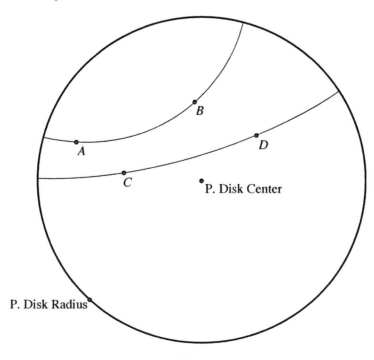

144

1. Open up a new sketch while keeping **Sketchpad: Samples: Sketches: Investigations: Poincare Disk.gsp** open in the background.

2. Choose **Custom Tool: Poincare Disk: Hyperbolic Line.** For the first line, you will need to place **P. Disk Center** and **P. Disk Radius**. The disk will appear. Then place two points inside the disk, and the hyperbolic line between them will appear. Label these points A and B.

3. Repeat (without placing the center or radius point for the disk) to draw the hyperbolic line CD. Drag points C and D so that these two lines do not intersect. Now AB and CD are considered to be parallel.

▷ **Exercise 2.** Draw two parallel hyperbolic lines AB and CD as above. Draw the hyperbolic line segment BC that intersects the two parallels. Measure the hyperbolic angles $\angle ABC$ and $\angle BCD$. What does this tell you about the Alternate Interior Angle relationship between parallel lines?

▷ **Exercise 3.** Draw two parallel hyperbolic lines AB and CD as above. Now construct another hyperbolic line CE through C. Adjust point E so that CE does not intersect AB. This shows that there are many lines through C parallel to AB in hyperbolic geometry.

Demonstration: Hyperbolic Circles [★ Hyperbolic Geometry.gsp: Hyperbolic circles]

A circle is defined to be the set of all points a given distance from a center point. However, distances are not what they appear to be in the Poincaré disk model.

1. Open up a new sketch while keeping **Sketchpad: Samples: Sketches: Investigations: Poincare Disk.gsp** open in the background.

2. Choose **Custom Tool: Poincare Disk: Hyperbolic Circle by CP.** You will need to place **P. Disk Center** and **P. Disk Radius**, and then the disk will appear. Place two points inside the disk, and the hyperbolic circle centered at the first point and passing through the second will appear. Label these points A and B.

3. Note that the circle does indeed look circular, but A doesn't look like the center.

4. Choose the **Custom Tool: Poincare Disk: Hyperbolic Distance** option. Select the point A and then B to measure the distance from A to B.

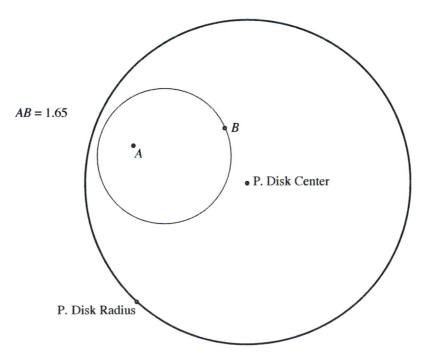

$AB = 1.65$

145

▷ **Exercise 4.** Draw a hyperbolic circle as in the previous demonstration, and measure the distance from the center A to point B on the circle. Select this circle, and use the **Construct: Point on Circle** command to place another point C on the circle. Measure the distance from the center A to C. Drag C around the circle; observe that the hyperbolic distance from A to C remains constant, so this is indeed a circle, at least in hyperbolic distances. Drag point B around to increase and decrease the radius, and observe that the distances remain equal: $AB = AC$.

▷ **Exercise 5.** Draw a hyperbolic circle centered at a point A that is close to the center of the Poincaré disk and passing through a point B. Measure the distance from A to B, and move B until the radius of this circle is close to 1. Draw another hyperbolic circle centered at a point X close to the rim of the disk. Adjust the radius of this circle so it too has radius 1. What does this say about the apparent areas of these circles?

Recall the construction of an equilateral triangle by ruler and compass: One begins with a line segment AB. Draw a circle centered at A that passes through B and then another circle centered at B that goes through A. The intersection of these two circles gives C, the third vertex of an equilateral triangle.

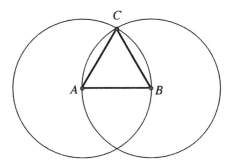

▷ **Exercise 6.** Draw an equilateral hyperbolic triangle, using the **Hyperbolic Segment** and **Hyperbolic Circle by CP** tools. Measure the hyperbolic distance along each edge to show that they are equal. Measure the three angles. Move your triangle around to different positions inside the disk. What do you notice about the angles? [★ Hyperbolic Geometry.gsp: Equilateral triangle]

▷ **Exercise 7.** Use the **Hyperbolic P. Bisector** tool to construct the perpendicular bisector of one of the sides of your triangle from Exercise 6. Note that the third vertex should fall on this line.

Demonstration: Right Triangles in Hyperbolic Geometry [★ Hyperbolic Geometry.gsp: Right triangle]

Construct a right triangle as follows.

1. Open up a new sketch while keeping **Sketchpad: Samples: Sketches: Investigations: Poincare Disk.gsp** open in the background.

2. Choose **Custom Tool: Poincare Disk: Hyperbolic Line.** You will need to place **P. Disk Center** and **P. Disk Radius**, and then the disk will appear. Place two points inside the disk, and the hyperbolic line through the two points will appear. Label these points A and B.

3. Use the **Hyperbolic Perpendicular** tool to drop a perpendicular to the line AB. You will need first to place a new point off the line, then select A and B. Label the new point C.

4. Place (using the **Point** tool or **Construct: Intersection**) and label point D at the intersection of AB and the perpendicular line.

5. Draw a hyperbolic line segment from A to C. We now have a right triangle $\triangle ADC$.

146

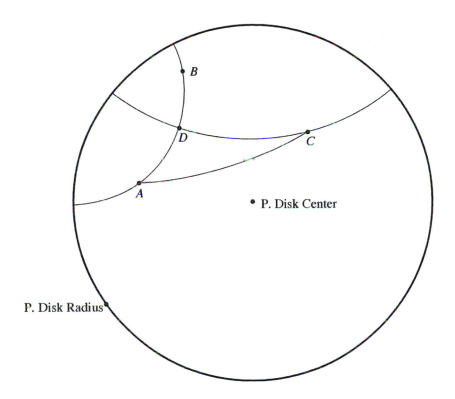

▷ **Exercise 8.** Draw a hyperbolic right triangle $\triangle ADC$ as above. Measure the distances AD, DC, and AC. Use **Measure: Calculate** to compute $AD^2 + DC^2$ and AC^2. What does this tell you about the Pythagorean theorem in hyperbolic geometry?

27. Curvature

Companion to Chapter 11.3 of <u>Symmetry, Shape, and Space</u>

In order to do tangent lines and osculating circles properly and precisely, you need a knowledge of calculus. However, you can gain intuition and find close but not exact measurements by using simple tools, such as circle templates or *The Geometer's Sketchpad*.

Recall the definition of the curvature of a curve at a point: It is the reciprocal of the radius of the osculating circle, $\kappa = \frac{1}{r}$. The sign of the curvature indicates on which side of the curve the osculating circle lies. For curves that represent the graphs of functions, this translates into positive if the circle lies above the curve and negative if it lies below the curve (or vice versa—it does not actually matter as long as you are consistent). For closed curves, things get a bit more complicated. Probably the best way to explain this is to imagine walking around the curve with one arm stuck out at right angles to the curve. If you reach one of the indicated points with your arm pointing toward the center of the osculating circle, then let the curvature be positive. If the center of the circle is on the opposite side of the curve from your arm, then let the curvature be negative. This obviously depends on which direction you choose to walk and which arm you stick out, but other choices will only reverse all of the signs. Again, consistency matters most of all. *The Geometer's Sketchpad* will compute the radius and the reciprocal of the radius for you, but you will have to keep track of the sign by hand.

Demonstration: Graphing Curves

To find the curvature of a curve, we must of course first graph the curve. *The Geometer's Sketchpad* has graphing tools provided under the **Graph** menu.

1. Choose **Graph: Grid Form: Square Grid** to draw a coordinate plane. You can adjust the point at position $(1, 0)$ to change the scale of the grid. However, this will change distances, and thus radii, and so the curvature.

2. To graph a curve on this grid, choose **Graph: Plot New Function**. We will give you the equation to enter here. In general, we will be graphing two types of curves: curves expressed as $f(x)$ (the default choice in the menu), and curves expressed as $r = f(\theta)$. To get this second option, go to the **Equation** button at the bottom of the **New Function** window. Hold this button down, and you will see the option $r = f(\theta)$. On the next page is the graph of the curve $f(x) = x^4 - 3x^2$.

148

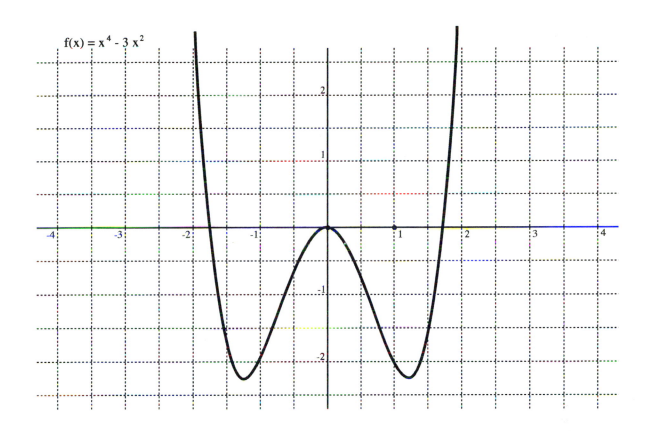

$f(x) = x^4 - 3x^2$

▷ **Exercise 1.** Use the **Graph: Plot New Function** command to graph the parabola $f(x) = x^2$. In the **New Function** window, you will enter this by clicking the buttons **x**, ∧, **2**, and then **OK**. The graph should appear on your grid.

Demonstration: Drawing Osculating Circles and Calculating Curvature

1. Once you have a curve with which to work, you can draw approximate osculating circles. Place a point on the curve: To do this, either use the **Point** tool or the **Construct: Point on Function Plot** command to place the point by eye, or use **Graph: Plot Point** to place the point precisely at a previously calculated position. Be sure that **Graph: Snap Points** is not checked.

2. Choose the **Circle** tool, and place the pointer approximately where you would guess the center of the circle of curvature to be. (You can adjust this later.) Hold the mouse button down while your move the cursor over to the point on the curve. When the point lights up, then release the mouse button. You should have a circle that goes through the point on the curve.

3. Select the center of the circle, and drag it until the circle fits the piece of curve near the point as closely as you can manage. This is your approximate osculating circle.

4. Select the circle, and use **Measure: Radius** to find its radius.

5. Choose the **Measure: Calculate** option. In the **Calculate** window that will appear, enter **1**, then ÷, and then click on your radius measurement. Now hit the **OK** button. The result is the approximate curvature (without the sign) of the curve at the selected point.

149

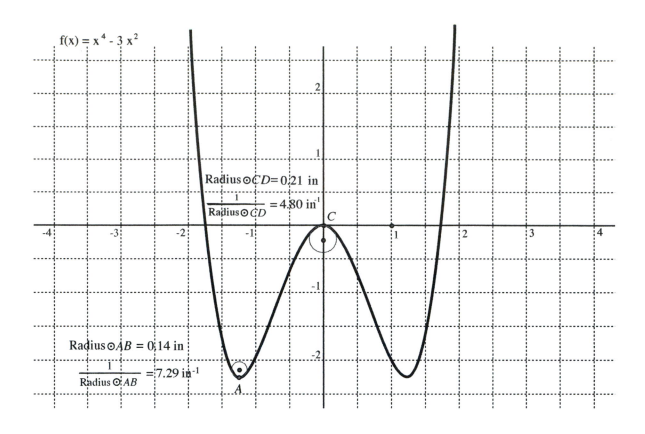

$f(x) = x^4 - 3x^2$

Radius $\odot CD = 0.21$ in

$\dfrac{1}{\text{Radius}\,\odot CD} = 4.80$ in^{-1}

Radius $\odot AB = 0.14$ in

$\dfrac{1}{\text{Radius}\,\odot AB} = 7.29$ in^{-1}

Note that by the sign conventions we have adopted, these computations imply that the curvature at A is $\kappa = 7.29$, while at C we have $\kappa = -4.80$.

For curves where you are asked to draw several osculating circles it helps to color code, so that the point on the curve, the osculating circle at the point, and the center of the osculating circle are all the same color. Use a different color for the curve itself and for each of the other osculating circles.

▷ **Exercise 2.** Draw the osculating circle for the curve $f(x) = x^2$ at the point $(0,0)$. Compute the approximate curvature at this point.

▷ **Exercise 3.** [SSS 11.3.2] Note that an ellipse is not described by a function since it fails the vertical line test. However, it can be written as two functions. Choose **Graph: Plot New Function**, and enter the function $f(x) = \sqrt{4 - \frac{4}{9}x^2}$. (You will have to first find the square root function under the **Functions** menu, where it is written as **sqrt**. Thus you will enter **sqrt** $(4 - (4 \div 9) * \mathbf{x} \wedge \mathbf{2})$ and then click **OK**.) On the same graph, choose **Graph: Plot New Function**, and graph $g(x) = -\sqrt{4 - \frac{4}{9}x^2}$. You should now see an ellipse. Use **Graph: Plot Points** to plot the points at positions $(3,0)$, $(-3,0)$, $(0,2)$, and $(0,-2)$. Now draw the osculating circles at these four points, and approximate the curvature at each point. [★ Curves.gsp: Ellipse]

▷ **Exercise 4.** [SSS 11.3.4] Using **Graph: Plot New Function**, enter the function $f(x) = \sin(x)$. (You will have to first find the sine function under the **Functions** menu, where it is written as **sin**.) Place points at the positions $(\frac{\pi}{2}, 1)$, $(-\frac{\pi}{2}, -1)$, $(-\frac{3\pi}{2}, 1)$, and $(\frac{5\pi}{4}, -\sqrt{2}/2)$ as shown. Now draw the osculating circles at these four points, and approximate the curvature at each point. Try drawing an osculating circle for the curve at the point $(0,0)$. What does this say about the curvature at $(0,0)$? [★ Curves.gsp: Sine]

150

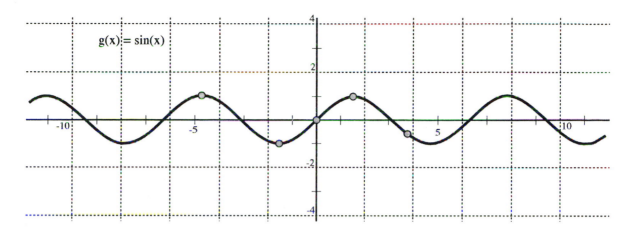

Some of the other shapes we are interested in are best plotted using polar functions. Since we will give the functions and instructions, these are no harder than cartesian functions. Polar functions are written in the form $r = f(\theta)$ where θ is an angle measure, usually between 0 and 2π.

▷ **Exercise 5.** **[SSS 11.3.6]** Choose **Graph: Grid Form: Square grid**. Now choose **Graph: Plot New Function**. In the **New Function** window that pops up, hold down the **Equation** button and slide down to $r = f(\theta)$. Now enter in the window the function $f(\theta) = \sqrt{\cos(2 * \theta)}$. You should see a lemniscate as pictured below. Use **Graph: Plot Points** to plot points on the curve at positions $(0.471, 0.333)$, $(-0.471, 0.333)$, $(0.471, -0.333)$, $(-0.471, -0.333)$, and $(-1, 0)$. Note that there is already a point at position $(1, 0)$ from the construction of the grid. Draw the osculating circles at these six points, using six different colors and color-coding the points and the centers of the circles the same colors as the circles with which they are associated. Compute the curvature at these six points. [★ Curves.gsp: Lemniscate]

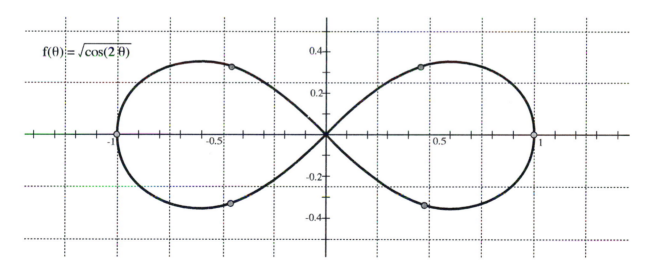

▷ **Exercise 6.** **[SSS 11.3.7]** Choose **Graph: Grid Form: Square grid**. Now choose **Graph: Plot New Function**. In the **New Function** window that pops up, hold down the **Equation** button and slide down to $r = f(\theta)$. Now enter in the window the function $f(\theta) = 2\cos(\theta) + 1$. You should see a limaçon as pictured on the next page. Use **Graph: Plot Points** to plot points on the curve at positions $(3, 0)$, $(0, 1)$, $(0, -1)$, $(1, 1.732)$, and $(1, -1.732)$. Note that there is already a point at position $(1, 0)$ from the construction of the grid. Draw the osculating circles at these six points, using six different colors and color-coding the points and the centers of the circles the same colors as the circles with which they are associated. Compute the curvature at these six points. [★ Curves.gsp: Limacon]

151

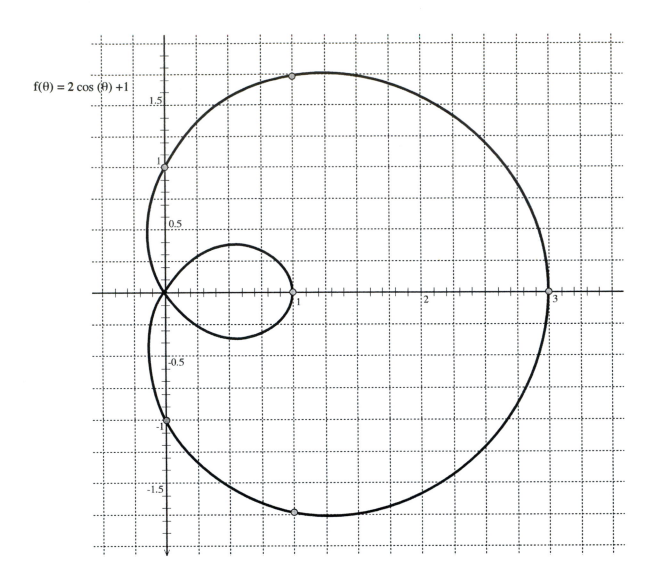

f(θ) = 2 cos (θ) +1

28. Soap Bubbles

Companion to Chapter 11.5 of <u>Symmetry, Shape, and Space</u>

The Geometer's Sketchpad does not model three-dimensional phenomena very well, so it isn't much use in studying the curvature of surfaces. However, it does an excellent job with the ruler and compass construction that gives the double bubble.

Demonstration: Double Bubble [★ Double Bubble.gsp]

1. Draw two line segments of length r_1 and r_2 which will represent the radii of your two bubbles. Park these in one corner of your sketch.
2. Place a point A on your sketch. Select this point and the line segment r_1, and choose **Construct: Circle by Center + Radius**.
3. Place a point D somewhere on the circle just constructed. This point will be one of the points of intersection of the two bubbles. Use **Construct: Ray** or the **Ray** tool to draw a ray from D to A.
4. Select D and choose **Transform: Mark Center** (or double-click on D). Select the ray DA, and use **Transform: Rotate** to rotate the ray by 60°. Rotate this new ray about D by another 60°.

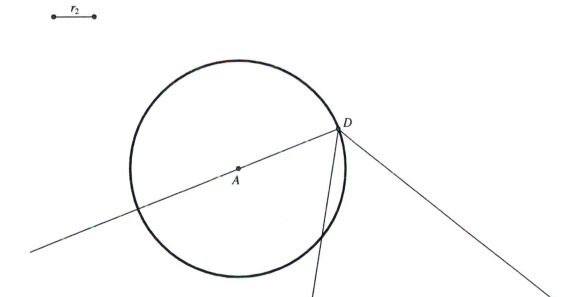

5. Draw a circle centered at D with radius r_2. Construct the intersection of this circle and the second ray, and label this point B. This will be the center of the second bubble.
6. Draw a ray from A through B.
7. Construct the intersection of the ray AB and the third ray constructed in Step 4, and label this point C. This point will be the center of the arc forming the wall between the bubbles.

153

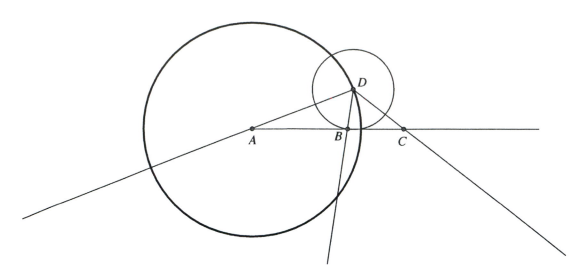

8. Use **Construct: Circle by Center + Radius** to draw a circle centered at B with radius r_1.

9. Use **Construct: Circle by Center + Point** to draw a circle centered at C and passing through D.

10. Construct the point E where the circles centered at A and B intersect.

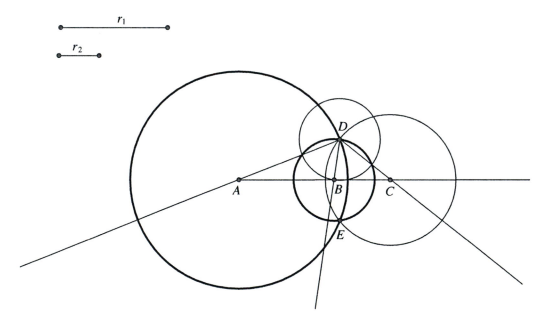

11. Select the circle centered at A, then the points D and E, and choose **Construct: Arc on Circle**. (Note that the arc is constructed counter-clockwise from the first point selected to the second: If you wanted the shorter arc, choose the circle, then E, and then D.)

12. Select the circle centered at B and then points E and D. Choose **Construct: Arc on Circle**.

13. Select the circle centered at C and then points D and E. Choose **Construct: Arc on Circle**.

14. Select all of the circles (but not the arcs), the rays, and the points, and hide them using **Display: Hide Objects**. Use **Display: Line Width: Thick** to thicken the arcs centered at A and B.

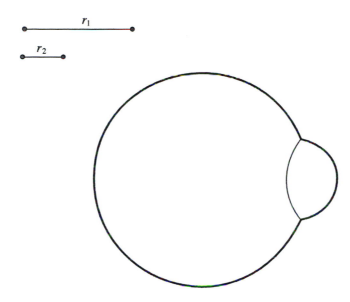

Drag the ends of the line segments that represent the radii to see the shape of the double bubble change. You can find the lengths of these radii using the **Measure: Length** command.

▷ **Exercise 1.** Draw a double bubble sketch as described in the previous demonstration. Draw the double bubble formed by two bubbles each with radius 1 inch.

▷ **Exercise 2.** [SSS 11.5.11] Draw a double bubble and the intermediate wall if one bubble has radius 2 inches and the other 1 inch.

▷ **Exercise 3.** [SSS 11.5.12] Draw a double bubble and the intermediate wall if one bubble has radius 1.5 inches and the other 1 inch.

29. Trees

Companion to Chapter 12.2 of <u>Symmetry, Shape, and Space</u>

A graph is a finite non-empty set of vertices and edges where each edge must have a vertex at both ends. A path is a string of edges, and a circuit is a path which begins and ends at the same vertex. A tree is a connected graph containing no circuits.

Drawing trees using *The Geometer's Sketchpad* is easy. Construct a starting vertex point. Then use the **Segment** tool, making sure that every new segment starts at an existing vertex but ends at a new vertex. That is, never connect a new branch to an existing one. Once a tree is drawn, moving the vertices and segments may change the way it looks, but it will not change the nature of the graph.

▷ **Exercise 1.** The following pictures are different representations of a single tree. Recreate the figures. Then drag the vertices in each picture until they all look the same.

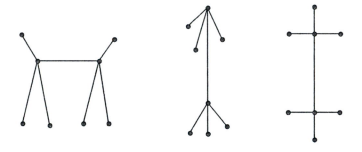

▷ **Exercise 2.** [SSS 12.2.3] Recreate the figures below. Then move the vertices on each tree until equivalent trees look the same. There are only four distinct trees.

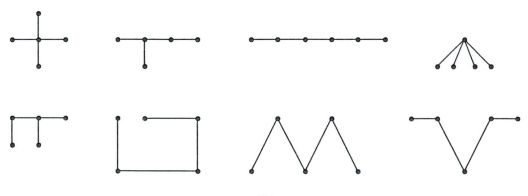

156

▷ **Exercise 3.** **[SSS 12.2.4]** Below are the numbers of combinatorially distinct trees for each number of vertices. Use *The Geometer's Sketchpad* to draw these trees.

Numbers of Vertices and Trees

Number of vertices	1	2	3	4	5	6	7	8
Number of trees	1	1	1	2	3	6	11	23

In some ways, circuits in a graph are redundant. They give you more than one way to get from point A to point B. In many other ways, they are useful. Building a road from A to B may be a good idea even if roads from A to C and from C to B already exist. However, if no roads exist and you are trying to connect three towns as efficiently as possible, the result will be a Steiner tree. The exercises below duplicate exercises from <u>Symmetry, Shape, and Space</u>.

▷ **Exercise 4.** **[SSS 12.2.6]** Using *The Geometer's Sketchpad*, put three points A, B, and C at the corners of an equilateral triangle. Put a fourth point D somewhere inside the triangle as shown below. Construct the line segments connecting each of the triangle vertices A, B, and C to the interior point D. Select each of these three segments in turn, and choose **Measure: Length**. Then choose **Measure: Calculate**, and click on **m$\overline{AD}$ + m$\overline{BD}$ + m$\overline{CD}$**. Clicking **OK** should put the sum on your sketch. Move D around until the total length is minimized. Next, use **Measure: Angle** to measure each of the angles $\angle ADB$, $\angle BDC$, and $\angle CDA$. What do you notice? [★ Trees.gsp: 1]

$m\ \overline{AD} = 1.15$ in
$m\ \overline{BD} = 0.88$ in
$m\ \overline{CD} = 1.19$ in

$m\ \overline{AD} + m\ \overline{BD} + m\ \overline{CD} = 3.22$ in

$m\ \angle ADB = 130.72°$
$m\ \angle BDC = 125.28°$
$m\ \angle CDA = 104.00°$

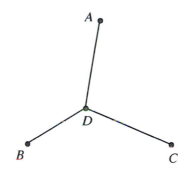

▷ **Exercise 5.** **[SSS 12.2.7]** Repeat the previous exercise using several nonequilateral but acute triangles $\triangle ABC$. What does the structure look like around D in each triangle when the sum of the distances is minimized?

▷ **Exercise 6.** **[SSS 12.2.9]** Repeat the previous exercise using several obtuse triangles. What can you say about the angles of the triangles that need a point other than one of the triangle vertices to minimize the "road distance" between towns?

▷ **Exercise 7.** [**SSS 12.2.10–12.2.12**] Construct vertices at the corners of three identical squares. Construct and measure the sum of the lengths of the line segments as in each of the pictures below. Move the interior vertices in the second and third pictures until the sum of the distances for each is minimized. Which picture has the smallest total distance required to connect the points, and at what angles do the interior line segments meet? [★ Trees.gsp: 2]

▷ **Exercise 8.** [**SSS 12.2.13**] Model the situation created by having four towns at random points in the plane. Discuss the Steiner trees you find to minimize the lengths of the trees connecting your towns. What angles between your towns give one or two Steiner vertices? Recall the acute versus certain obtuse triangular setups above. Can you find more than one Steiner tree for some setups? Try the following four layouts.

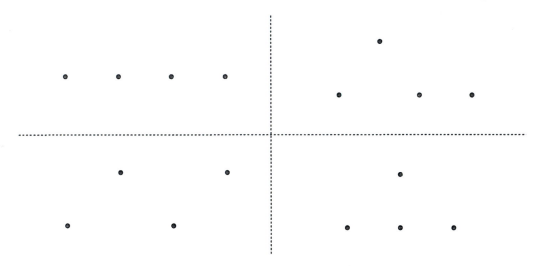

30. Graphs and Mazes

Companion to Chapters 12.1 and 12.3 of <u>Symmetry, Shape, and Space</u>

The Geometer's Sketchpad is an excellent tool for drawing many graphs and mazes. The trees in the previous section are—in some ways—the simplest graphs and the easiest to construct. Complete and complete bipartite graphs are most commonly drawn using only straight edges and so are very easy to construct. Planar graphs are more difficult to identify, and therefore drawings using *The Geometer's Sketchpad* are not as helpful. Determining whether a graph is planar requires trying to redraw edges so that they do not overlap. This often requires drawing an arc for an edge instead of a straight line. For the same reason, the software does not work as well for graphs with more than one edge between two vertices, though arcs of carefully constructed circles can be used.

One strategy for calculating the Euler characteristic of a graph involves removing edges, as few as possible, from the graph until a tree is left. Determining the relationship between this number of edges and the Euler characteristic is Exercise 12.2.2 of <u>Symmetry, Shape, and Space</u>. *The Geometer's Sketchpad* is useful for finding the number of edges to remove. Assuming you can draw the original graph, select edges and hide them until only a tree remains.

A complete graph is a graph in which every pair of distinct vertices is joined by exactly one edge. The complete graph on n vertices is denoted by K_n. K_5 is pictured below.

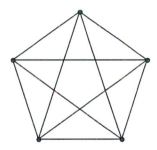

▷ **Exercise 1.** [SSS 12.1.16] Draw K_n for $n = 1, 2, 3, 4, 6$, and 7.

A complete bipartite graph has the vertices divided into two groups. There is exactly one edge between each pair of vertices from different groups. There are no edges between pairs of vertices from the same group. The complete bipartite graph with n vertices in the first group and m vertices in the second group is denoted by $K_{n,m}$. $K_{2,5}$ is pictured below.

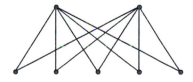

▷ **Exercise 2.** [SSS 12.1.17] Draw $K_{3,2}$, $K_{3,4}$, $K_{3,5}$, $K_{4,5}$, $K_{5,5}$, and $K_{6,7}$.

Many mazes are constructed on grids. Therefore, the **Graph: Grid Form** command provides an excellent medium for creating mazes. The **Graph: Snap Points** option ensures uniformity of path size, and the **Display: Hide** command allows a final picture free of construction objects. The **Construct: Interior** command can be used to shade dead ends. The **Segment** tool and **Edit: Undo** can be used to solve a maze via trial and error. The **Segment** tool can also be used to draw the graph of a maze: Draw a line segment down each path, and then hide the maze or copy the graph to another position. The graph can then be simplified by moving or deleting edges.

The simplest mazes are based on trees. In such a maze, there is only one way to get to each cell.

Demonstration: Construct a Maze Based on a Tree [★ Mazes.gsp: 1–4]

1. Open a sketch and choose **Graph: Grid Form: Square Grid**. Choose **Graph: Snap Points** to ensure the maze is uniform.

2. Outline a 10 × 10 section of the grid as your working space using the **Segment** tool. You may want to choose **Display: Line Width: Wide** and **Display: Color** to change the color of your maze and make it easier to see. Leave a gap one square wide along the border of your outline for an entrance and another for the exit.

3. Construct trees of line segments along the grid lines but always starting on the external border. Remember that a tree can branch but cannot reconnect to itself.

4. There must be a segment connected to every vertex inside the working area to complete the maze. Then choose **Graph: Hide Grid**.

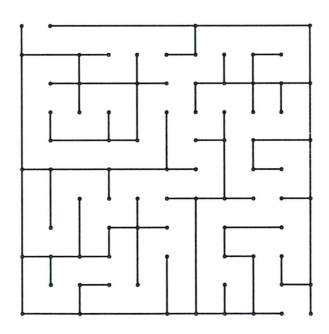

5. Now we will confirm that the maze is based on a tree. Choose **Graph: Snap Points** a second time to turn off that option.

6. Use the **Segment** tool and a different color to trace all the paths of the maze.

7. Hide the maze and adjust the remaining graph of paths until it is clearly a tree.

160

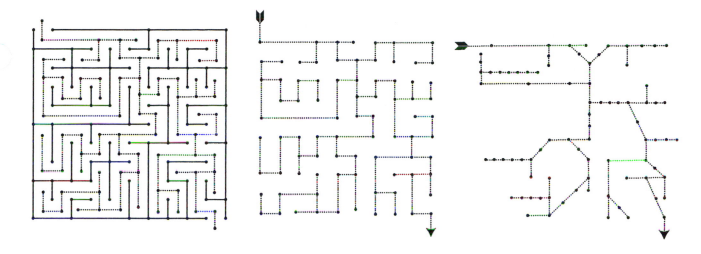

▷ **Exercise 3.** [SSS 12.3.2] Using a 10 × 10 section of a square grid, draw a different maze based on a tree.

▷ **Exercise 4.** [SSS 12.3.1] Make a copy of your maze and turn off the **Snap Points** command. Using the **Segment** tool and a different color for the lines, trace all the paths in your maze, hide the maze, and confirm that a tree is formed.

A *unicursal* maze, or *labyrinth*, is a maze with only one path.

▷ **Exercise 5.** [SSS 12.3.3] Construct a labyrinth on a 10 × 10 section of a square grid. Hide the grid in your final solution.

▷ **Exercise 6.** Draw a maze on a 10 × 10 section of a square grid where at least one of the trees forming the inside edges of the maze is not connected to the external border.

▷ **Exercise 7.** Draw the graph formed by the paths of the maze constructed in Exercise 6. Simplify the graph until it is clear that the graph is not a tree.

One method for solving a maze involves shading dead ends. By selecting the vertices around a dead end and choosing **Construct: Polygon Interior**, you can color in parts of the maze until the correct path is clear. [★ Mazes.gsp: 5–6]

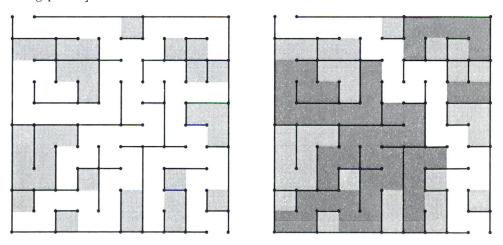

▷ **Exercise 8.** Solve the maze you constructed in Exercise 3 by shading dead ends.

161

THE GEOMETER'S SKETCHPAD

Quick Reference
For Windows® and Macintosh®
Version 4.0

About Document Windows

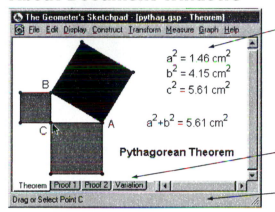

Sketch Plane. Draw new objects here using the **Point, Compass, Straightedge,** and **Text** tools. Drag objects to explore relationships using the **Selection Arrow** tool. Select objects and use menus to reformat or measure them, or to construct new objects defined by selected objects.

Page Tabs. In multi-page documents, use tabs to switch pages. (To add new pages, choose **File | Document Options.**)

Status Line. Describes current selections or tool action.

About the Toolbox

Selection Arrow and **Translate** tool: Click on objects in sketch to select them. Drag objects to move (**Translate**) them. (Press icon to pull out **Translate, Rotate,** and **Dilate** arrows.)

Point tool: Click in blank sketch area to create an independent point. Or click on object to create a point on that object.

Compass (Circle) tool: Press mouse button to create center, drag to create circle, release to create radius control point. Center and radius points can be independent points or points on objects.

Straightedge (Segment) tool: Press to create first endpoint, drag to create segment, release to create second endpoint. (Press icon to pull out **Segment, Ray,** and **Line** tools.)

Text tool: Double-click in blank area to create caption. Click on object to display or hide label. Drag label to reposition. Double-click on label, measure, or caption to edit or change style.

Custom tools: Press icon to display commands for creating new tools, and a list of all available custom tools. Choose custom tool from list to use in sketch. (See **Custom Tools.**)

Selecting Objects

Many menu commands require that you first select objects in your sketch to act upon. Commands that are gray are unavailable. Make unavailable commands available by first selecting their necessary objects.

To select one or more objects	Click each unselected object with **Selection Arrow.**
To select objects with selection rectangle	Click in blank sketch area with **Selection Arrow;** drag to define rectangular selection area; release to select all objects in, or partially in, selection rectangle.
To deselect one or more objects from group	Click each selected object with **Selection Arrow.**
To deselect all objects	Click in blank sketch area with **Selection Arrow.**

Dragging and Animating Objects

Dynamic Geometry® puts mathematics into motion. In Sketchpad™, objects move according to their mathematical relationship to the objects you drag. Use dragging to examine related cases, explore properties, and form conjectures. Use animation for more complex motions or for mathematical presentations.

Dragging with the Arrow Tools

Press and hold **Selection Arrow** to open pop-up arrow palette. Drag right to choose **Translate**, **Rotate**, or **Dilate** arrows. Dragging an unselected object with an arrow drags only that object. Dragging a selected object drags that object and all other selected objects. Use the **Translate** arrow to slide objects without turning. **Rotate** arrow turns objects around a marked center point. **Dilate** arrow shrinks or enlarges objects about the marked center.

Animating Objects with the Motion Controller

Choose **Display | Show Motion Controller** to show the Motion Controller. Motion Controller targets the objects currently selected in your sketch.

Target. Displays current (selected) target of Motion Controller actions. When objects are animating, press and hold for list of all moving objects.

Animate Button. Begins animating selected objects.

Reverse Button. Changes direction of selected (or all) objects animated along specific paths.

Stop Button. Stops animating selected objects, or stops all animated objects if no objects are selected.

Pause Button. Temporarily stops all animations. Press again to resume.

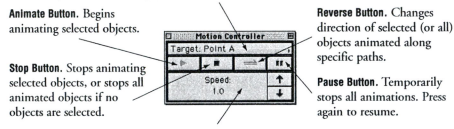

Speed Controls. Changes speed of selected animated objects, or all animated objects if no objects are selected. Click speed number and type to change speed numerically; or use arrows to increase or decrease present speed.

Animated Objects	How They Move
Independent points—drawn with **Point** tool	At random in the plane
Points on paths (segments, lines, circles, etc.) constructed with **Point** tool or by choosing **Construct	Point On Object**	Continuously along or around their path. Use Animate action buttons for control over initial direction, speed, etc.
Parameters—created by **Graph	New Parameter**	Continuously or discretely over some numeric domain, set in the Parameter's Properties dialog box
All other objects	By animating the objects ("parents") that define them

Using Action Buttons to Animate or Move Objects

Select any objects and choose **Edit | Action Buttons | Animation** to create an action button that animates those objects continuously. (Action buttons offer more options than the Motion Controller.) Select two points and **Edit | Action Buttons | Movement** to create a button that moves the first point to the second point, then stops. Other action buttons allow you to show and hide objects, switch document pages, link to web pages, or create presentations by combining the actions of other buttons.

Constructing and Transforming Objects

Use the Construct menu to define new geometric objects based on existing objects. Use the Transform menu to construct the translated, rotated, dilated, or reflected images of existing objects.

Using the Construct Menu

All Construct menu commands require specific selections. If a menu command you wish to use appears gray, check that you have selected the required objects (and only those objects) in your sketch. The Status Line (at the bottom of the sketch) describes current selections.

To Construct:	Select:
Point On Object	1 or more paths (segments, rays, lines, circles, arcs, interiors, axes, function plots, or point loci)
Midpoint	1 or more segments
Intersection	2 straight objects (segments, rays, etc.), circles, or arcs
Segment, Ray, Line	2 or more points
Parallel Line, Perpendicular Line	1 point and 1 or more straight objects; or 1 straight object and 1 or more points
Angle Bisector	3 points (select vertex of angle to bisect second)
Circle By Center+Point	2 points (select center of circle first)
Circle By Center+Radius	1 point and 1 segment or distance measurement
Arc On Circle	1 circle and 2 points on circumference
Arc Through 3 Points	3 points
Interior	3 or more points for **Polygon Interior**; 1 or more circles for **Circle Interior**; one or more arcs for **Arc Segment** or **Arc Sector Interior**
Locus	1 driver point (constructed on a path) and 1 driven object (point, segment, circle, etc.) whose position depends on driver point

Using the Transform Menu

Use the basic Transform commands—**Translate, Rotate, Dilate,** and **Reflect**—to construct the transformed image of one or more selected geometric object. Use **Mark** commands to specify other objects as transformational parameters. For example, **Mark Center** marks a selected point as center of future rotations and dilations. Marked parameters remain marked until you mark new ones. Shortcuts: Double-click a point to **Mark Center**; double-click a straight object to **Mark Mirror**.

Mark last selected point as center for rotation or dilation.
Mark last selected segment, ray, line, or axis as mirror for reflection.
Mark last 3 selected points or measurement as angle for rotation.
Mark 2 selected segments, 3 collinear points, or measurement as ratio of dilation.
Mark last 2 selected points as initial and terminal points for translation vector.
Mark selected measurement as distance for translation.

Translate selected objects by fixed or marked polar or rectangular vector.
Rotate selected objects by fixed or marked angle around marked center point.
Shrink or stretch selected objects by fixed or marked ratio about marked center.
Reflect selected objects across line, ray, segment, or axis marked as mirror.

Iterate a construction based on selected independent points and parameters.

Transform
Mark Center
Mark Mirror
Mark Angle
Mark Ratio
Mark Vector
Mark Distance

Translate…
Rotate…
Dilate…
Reflect…

Iterate…

Working with Measurements, Calculations, and Functions

Use the Measure menu to measure geometric and analytic properties of selected objects, and to create calculations that express or explore relationships between measured values. Use Graph menu commands to create coordinate systems and define, plot, and differentiate functions.

Measuring Properties

Measure menu commands require specific selections. If a menu command appears gray, check that you have selected the required objects (and only those objects) in your sketch. The Status Line (at the bottom of the sketch) describes the current selections.

Measurement Tips

Drag measured objects to change a measurement's value.

Double-click a measurement with the **Text** tool to change its name.

Choose **Edit | Properties | Value** to change a measurement's precision (number of displayed digits).

Choose **Edit | Preferences | Units** to change the units of all measured angles and distances in the sketch.

To Measure:	Select:
Length	1 or more segments
Distance	2 points; or 1 point and 1 straight object
Perimeter	1 or more polygon or arc interiors
Circumference	1 or more circles or circle interiors
Angle	3 points (select vertex of angle second)
Area	1 or more circles or interiors
Arc Angle, Arc Length	1 or more arcs
Radius	1 or more circles or arcs
Ratio	2 segments, or 3 collinear points
Coordinates	1 or more points
Abscissa, Ordinate	1 or more points
Coordinate Distance	2 points
Slope	1 or more straight objects
Equation	1 or more lines or circles

New Function

$$f(x) = \frac{a \sin(b \cdot x^2)}{2x}$$

`( a * sin( b * x^2)) / (2x)`

7	8	9	+	^	Values
4	5	6	−	(	Functions
1	2	3	*	)	Units
0	.	x	÷	←	Equation

Help   Cancel   OK

The Calculator appears when you choose **Measure | Calculate** to create a new calculation or either **Graph | New Function** or **Graph | Plot New Function** to create new functions.

Build an expression for your calculation or function using the numbers and operators on the keypad, your computer keyboard, and the Values, Functions, and Units pop-up menus.

Click on measurements and functions in your sketch to add them to the expression.

Use the Equation menu to choose between $x = f(y)$, $y = f(x)$, $\theta = f(r)$, and $r = f(\theta)$. (Applies to functions only.)

Click **OK** when expression is complete. To change existing expressions, select a calculation or function and choose **Edit | Edit Definition**. Shortcut: Double-click calculation or function.

Working with Function Plots

Graph | Plot Function plots selected functions on the current coordinate system. Drag arrowhead endpoints or use **Plot Properties** to change the domain of the plot. Double-click original function (not plot) to change the plotted equation or form (rectangular or polar). Construct points on function plots with **Point** tool or **Construct | Point On Object**.

Working with Coordinate Systems

Graph | Define Coordinate System creates a new coordinate system. Drag unit point(s) or numbers on axis tick marks to change scale. **Graph | Grid Form** switches active coordinate system between square, rectangular, and polar grids.

Formatting Objects

Use the Display menu to change the appearance or movement of selected objects. If a menu command you wish to use appears gray, check that you have selected appropriate objects for that command.

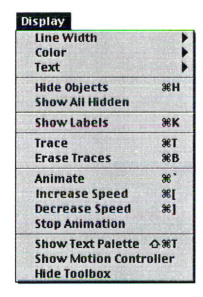

Display selected objects with thick, thin, or dashed lines.
Change the color of selected objects.
Change font for selected captions, measurements, or labels.

Hide selected objects (without affecting their role in definitions).
Show all previously hidden objects.

Show labels of selected objects. (Alternates to **Hide Labels**.)

Leave trace of selected objects when moved (checked/unchecked).
Remove any visible traces from window.

Animate selected objects (see **Motion Controller**).
Make active or selected animations faster (see **Motion Controller**).
Make active or selected animations slower (see **Motion Controller**).
Stop active or selected animations (see **Motion Controller**).

Show (or hide) Text Palette (see **Text Palette**).
Show (or hide) Motion Controller (see **Motion Controller**).
Hide (or show) Toolbox (see **Toolbox**).

Using the Text Palette

The Text Palette describes the text style of selected objects' labels, as well as the style of selected captions, measurements, and functions. Change the Text Palette settings to affect selected text, or use Symbolic Notation tools to add mathematical formatting as you edit captions with the **Text** tool. Note: For geometric objects with visible labels, use the Text Palette to change the color of the selected objects' *labels*; use **Display | Color** to change the color of the *objects* themselves.

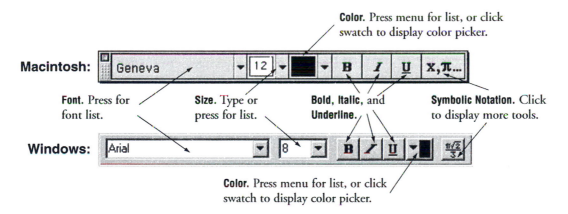

Setting Other Object Properties

Choose **Edit | Properties** with a single selected object to change Object, Label, Value, Plot, Parameter, Button, or Iteration properties. (Different panels appear for different types of objects.) Shortcut: Right-click an object (Windows) or Ctrl+click an object (Macintosh) to access its **Properties**.

Working with Custom Tools

Custom tools are tools you create yourself based on example constructions. Custom tools reside in the document in which you create them (unless you move them with **Tool Options**). Use custom tools like other Toolbox tools after choosing them from the Custom Tools menu in the Toolbox.

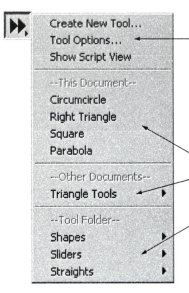

Select an entire construction and choose **Create New Tool** to make a tool that duplicates your construction when used in any document. Use **Tool Options** to rename custom tools, remove them from the active document, or copy them between documents.

Choose **Show Script View** to display a readable description of the most-recently chosen custom tool, and to modify the tool. (Alternates to **Hide Script View**.)

Choose any tool from the rest of the menu to use in your sketch.

These custom tools are stored in the active document.

These custom tools are stored in other open documents. Tools in any open document can be used in the active document.

These custom tools are stored in documents in the Tool Folder (next to the Sketchpad application on your hard disk). Tools stored in the Tool Folder when Sketchpad starts are always available.

Esc Key

The Esc key "escapes" from special states. Press Esc repeatedly to:

- Stop caption editing
- Choose the **Selection Arrow** tool
- Deselect all objects
- Stop all animations
- Erase all traces

Context Menu

The Context menu displays commonly used commands for a given object or document. Right-click (Windows) or Ctrl-click (Macintosh) an object to display that object's Context menu. Right-click (Windows) or Ctrl-click (Macintosh) in a document's blank space to display that document's Context menu.

More Help

The Help menu presents an electronic version of *The Geometer's Sketchpad Reference Manual* in your web browser—consult it for more information. Also, browse example Sketchpad documents installed in the **Samples** folder; and visit the Sketchpad Resource Center at http://www.keypress.com/sketchpad/.

References

Dan Bennett, <u>Exploring Geometry with The Geometer's Sketchpad</u>, Key Curriculum Press, Emeryville CA, 1999.

Steven Chanan, ed., <u>101 Project Ideas for The Geometer's Sketchpad</u>, Key Curriculum Press, Emeryville CA, 2000.

Steven Chanan, <u>The Geometer's Sketchpad: Learning Guide</u>, Key Curriculum Press, Emeryville CA, 2001.

L. Christine Kinsey and Teresa E. Moore, <u>Symmetry, Shape, and Space</u>, Key College Publishing, Emeryville CA, 2002.

Cathi Sanders, <u>Perspective Drawing with The Geometer's Sketchpad</u>, Key Curriculum Press, Emeryville CA, 1994.

Scott Steketee, Nicholas Jackiw, Steven Chanan, <u>The Geometer's Sketchpad: Reference Manual</u>, Key Curriculum Press, Emeryville CA, 2001.

David Thomas, <u>Active Geometry</u>, Brooks/Cole Publishing, Pacific Grove CA, 1998.